Current Trends in Wheat Research

Edited by Mahmood-ur-Rahman Ansari

Published in London, United Kingdom

IntechOpen

Supporting open minds since 2005

Current Trends in Wheat Research
http://dx.doi.org/10.5772/intechopen.87473
Edited by Mahmood-ur-Rahman Ansari

Contributors
Juan Manuel Sanchez-Yañez, Arif Hasan Khan Robin, Afsana Hannan, Md. Najmol Hoque, Lutful Hassan, Mirza Abdul Qayyum, Shafqat Saeed, Unsar Naeemullah, Muhammad Wajid, Aboubakar Siddique, Amar Matloob, Rumail Shahid, Hafiz Ubaid Ur Rehman Zia, Huda Bilal, Muhammad Ramzan, Astrid Jankielsohn, Harmeet Singh Singh Bakala, Loveleen Kaur Sarao, Puja Srivastava, Ankita, Kamalpreet Singh Mandahal, Muhammad Imran, Orhan Özçatalbaş, Leandro José Dallagnol, Keilor da Rosa Dorneles, Andrea Elizabeth Román Ramos, Olayinka Bolaji Umar, Lawal Amudalat Ranti, Abdulra'uf Lukman Bola, Abdulkareem Khadijat Abdulhamid, Abdulbaki Shehu Abdulbaki, Murtadha Ramat Biola, Kayode Oluwagbenga Victor, Mahmood-ur-Rahman Ansari, Nazia Naheed, Parwsha Zaib, Tayyaba Shaheen, Kanval Shaukat, Akmaral U. Issayeva

First published in London, United Kingdom, 2022 by IntechOpen
IntechOpen is the global imprint of INTECHOPEN LIMITED, registered in England and Wales, registration number: 11086078, 5 Princes Gate Court, London, SW7 2QJ, United Kingdom
Printed in Croatia

British Library Cataloguing-in-Publication Data
A catalogue record for this book is available from the British Library

Additional hard and PDF copies can be obtained from orders@intechopen.com

Current Trends in Wheat Research
Edited by Mahmood-ur-Rahman Ansari
p. cm.
Print ISBN 978-1-83968-593-4
Online ISBN 978-1-83968-594-1
eBook (PDF) ISBN 978-1-83968-595-8

For EU product safety concerns:
IN TECH d.o.o., Prolaz Marije Krucifikse Kozulić 3, 51000 Rijeka, Croatia, info@intechopen.com or visit our website at intechopen.com.

Meet the editor

Mahmood-ur-Rahman Ansari, Ph.D., is an Associate Professor of Molecular Biology at the Department of Bioinformatics and Biotechnology, GC University – Faisalabad, Pakistan. He received his BSc (Hons) in Plant Breeding and Genetics from the University of Agriculture, Faisalabad, Pakistan, in 2003. He obtained an MPhil and Ph.D. in Molecular Biology from the National Centre of Excellence in Molecular Biology, Lahore, Pakistan, in 2006 and 2011, respectively. He has published more than eighty papers in international peer-reviewed journals in the fields of molecular biology, biotechnology, and bioinformatics. He has also published more than fifteen book chapters and edited four books. His research interest is to understand the molecular mechanisms of stress tolerance in plants. He has been involved in the genetic modification of rice and cotton, their greenhouse and field testing, as well as biosafety studies. He is a member of various national and international professional societies.

Contents

Preface

Wheat is an important crop and is a staple food in almost more than half of the world. Currently, it feeds about 40% of the global population and contributes 20% in terms of total calories and protein intake. The present production does not meet the demands of the growing population worldwide, especially in developing countries. The alarming gap between increasing demand and current production is a big challenge for scientists. To meet this challenge, either area under production or yield per unit area or both need to be increased. The increasing area seems almost impossible because of constraints like drought, salinity, water logging, and trends in urbanization. Increasing yield, on the other hand, is a potential option that is possible through adopting better management practices and advanced technologies. This book discusses strategies to produce wheat under biotic and abiotic stress conditions. Developing biotic and abiotic stress tolerant wheat plants may lead to better production. Different sections of the book discuss diverse aspects of *Current Trends in Wheat Research* with special reference to biotic and abiotic stresses. This book contains three sections including an introductory chapter in the first section that provides an overview of recent approaches for the improvement of wheat.

Section 2 discusses the status of innovations in biotic stress tolerance in wheat. It contains chapters discussing various diseases of wheat as well as insect pests of wheat crop. It also describes strategies to develop better plants having biotic stress tolerance. Section 3 discusses the latest information regarding various approaches to developing abiotic stress-tolerant wheat plants. The chapters in this section contain useful information on the drought tolerance phenomenon and water use efficiency in wheat plants. The information in this section is especially important in the context of the current climate change scenario.

The information presented in this book is of great importance for research scholars, researchers, academicians, and the general public as well as other stakeholders. I would like to thank IntechOpen for giving me an opportunity to edit this book. I am also thankful to Author Service Manager Mr. Josip Knapic for his valuable help throughout the editing process. I must also thank my research student Ms. Parwsha Zaib for her assistance in the preparation of the introductory chapter. I am especially thankful to all the authors for their valuable contributions.

Mahmood-ur-Rahman Ansari, Ph.D.
Department of Bioinformatics and Biotechnology,
Government College University,
Faisalabad, Pakistan

Section 1

Introduction

Chapter 1

Introductory Chapter: Current Trends in Wheat Research

Nazia Nahid, Parwsha Zaib, Tayyaba Shaheen, Kanval Shaukat, Akmaral U. Issayeva and Mahmood-ur-Rahman Ansari

1. Introduction

Wheat (*Triticum aestivum*) is known as one of the most important cereal crops and is extensively grown worldwide [1]. Wheat contributes to 50% and 30% of the global grain trade and production respectively [2]. Wheat is also known as a staple food in more than 40 countries of the world. Wheat provides 82% of basic calories and 85% of proteins to the world population [3, 4]. Wheat-based food is rich in fiber contents than meat-based food. Dough produced from bread wheat flour has different viscoelastic properties than other cereals. It is considered a higher fiber food. Therefore, its positive effects on controlling cholesterol, glucose, and intestinal functions in the body were observed [5]. Primarily, wheat is being used to make *Chapatti* (Bread) but it also contributes to other bakery products. Wheat utility and high nutritional value made it the staple food for more than 1/3rd population of the world. Wheat grain is separated from the chaff and stalks after the harvesting of wheat. Stalks of wheat are further used in animal bedding and construction material. Globally, the need for wheat production is enhancing even in countries having unfavorable climates for its production. Global climate changes are badly affecting the production of wheat and it raised the concern for food security.

It is estimated that annual cereal production should be increased by 1 billion tons to feed the expected population of 9.1 billion by 2050 [1]. The current scenario demands an increase in crop productivity to meet the increased requirements of food supply [6]. Wheat is grown in tropical and subtropical regions which experiences a lot of stress. These stresses result in a reduction of yield [7]. Major environmental stresses include cold, salinity, heat, and drought which are drastically affecting its yield. However, water and heat are considered as the key environmental stresses which caused in reduction of the wheat yield globally [8, 9]. So, genetic improvements related to yield and stress tolerance are mandatory to enhance the production of wheat [10, 11].

2. Genetically modified wheat plants

Genetically modified wheat plants have been produced by the use of bacteria. Wheat plants were inoculated with the plant-growth-promoting bacteria (PGPB) which resulted in the higher expression of abiotic stress (mainly drought and salinity) tolerant genes [12]. PGPB inoculated wheat cultivars also showed the higher expression of genes encoding antioxidant-enzymes, such as *catalase* (CAT),

IntechOpen

S. No.	Gene Name	Trait/Phenotype	Reference
1.	*ZmDof1*	Increased yield	[14]
2.	*TaNFY-A-B1*	Increased Nitrogen and Phosphorus uptake	[15]
3.	*TaNF-YB4*	More grain yield	[16]
4.	*TaNAC2-5A*	More root growth	[17]
5.	*OsSS-I*	Increased heat tolerance	[18]
6.	*TaGS2*	More yield	[19]
7.	*ZmAGPase*	More yield	[20]
8.	*NtNR*	More yield, More seed protein contents	[21]
9.	*TaVIT2*	Iron biofortification	[22]
10.	*HD-ZipI*	Drought and frost tolerance	[23]
11.	*DREB1A*	Drought tolerant	[24]
12.	*CspA & CspB*	Drought-stress tolerance	[25]
13.	*HaHB4*	Abiotic-stress tolerance	[26]
14.	*TaDREB3*	Drought-stress tolerance	[27]

Table 1.
Development of transgenic wheat having various traits/phenotypes.

peroxidase, *ascorbate peroxidase* (APX), and *glutathione peroxidase* (GPX). So, it was concluded that PGPB used in wheat plants resulted in increased tolerance to abiotic stresses [12]. Cold shock proteins increase the survival of bacteria in severe environmental conditions. *CspA* and *CspB* genes from bacteria were transformed into wheat. Transgenic wheat plants expressing *SeCspA* and *SeCspB* were observed to have decreased water loss rate, increased proline and chlorophyll contents under salinity, and less water-stress conditions [13]. It was further investigated that *SeCsp* transgenic wheat plants resulted in enhanced weight and yield of grain than the control plants. *SeCspA* transgenic wheat plants were observed to have an improved water-stress tolerance than the control plants (**Table 1**, [13]).

Gluten is a protein comprised of gliadins found in wheat. Gluten is the main cause of coeliac disease in individuals. Bread-making quality of wheat is determined by the gluten proteins. Wheat varieties with less gliadin contents were produced using gene-editing technologies and RNAi (RNA interference). Wheat lines lacking immunogenic gluten were produced. Low immunogenic gluten and more nutritional values were added in one wheat line named E82. A better microbiota profile (protection microorganisms available in the gut) was observed in the NCWS patients using the bread made with E82 [28]. Plant cuticle has a positive role in the protection of plant against biotic and abiotic stresses. Wheat plants transformed with *TaSHN1* resulted in increased water-stress tolerance by reducing the leaf stomatal density and changing the composition of the cuticle [29].

3. Biotic stress tolerance in wheat

Wheat is considered an excessive contributor toward the human calorie intake [30]. Pests and pathogens cause yield losses in wheat up to 21.5% of the total losses and could be reached to 28.1% [31]. Wheat is affected by the fungal disease, powdery mildew caused by *Blumeria graminis* f. sp. tritici (Bgt). Powdery mildew is a damaging disease that resulted in greater loss of wheat [32]. Broad-spectrum resistant genes (BSR) are considered to have the most significant role to control powdery mildew. *CMPG1-V* gene was cloned from the *Hynaldia villosa* and it was observed that higher expression of *CMPG1-V* gene resulted in the Broad-spectrum resistance against powdery mildew [33, 34]. Barley *chi26* gene could also be used to enhance the resistance against powdery mildew and rust through genetic modification [35]. Some epigenetic regulators were determined to have a role in wheat powdery mildew resistance. *TaHDT701* is a histone deacetylase that was found as a negative regulator of wheat defense against powdery mildew. *TaHDT701* was observed to be associated with the one repeat protein (TaHOS15) and RPD3 type histone deacetylase TaHDA6. Knockdown of this histone deacetylase complex (*TaHDT701, TaHDA6, TaHOS15*) in wheat resulted in increased powdery mildew tolerance [36].

Fusarium graminearum is a plant fungal pathogen that causes a devastating disease called Fusarium head blight in wheat. It results in the reduction of wheat production. Genetic techniques were used to increase the FHB (Fusarium head blight) resistance in wheat. Transgenic wheat plants expressing barley class II chitinase gene 2 were observed to have a higher resistance against *Fusarium graminearum* [37]. *Lr10* and *Lr21* were cloned and transformed into wheat. The transgenic plants were reported to be resistant to leaf rust disease. Evolution and diversification of *HIPPs* (heavy metal-associated isoprenylated plant proteins) genes were studied in Triticeae [38]. *HIPPs* genes of *Hynaldia villosa* were cloned through homology-based cloning. Transgenic wheat having *HIPP1-V* was developed and the role of *HIPP1-V* in cadmium stress was characterized. It was observed that higher expression of this gene resulted in increased tolerance to cadmium stress. Therefore, *HIPP1-V* could be used to increase the tolerance in wheat against cadmium [39].

4. Abiotic stress tolerance in wheat

Grain number, weight, and size are greatly reduced under the negative effects of environmental stresses. However, the timing, duration, and intensity of stress determine the severity of the negative effects [40, 41]. Wheat is a major source of protein and calories for the human diet. High temperature is badly affecting the yield of wheat which is a main concern worldwide. Drought and heat stresses are the two main abiotic stresses which are playing a greater role in the reduction of wheat yield. Reduction in starch contents, photosynthetic activity, grain number, and chlorophyll contents in the endosperm is caused due to rise in temperature. Heat stress results in the accumulation of reactive oxygen species (ROS) which is the main reason for higher oxidative damage to the plant. Heat stress also results in the variation of wheat biochemistry, morphology, and physiology. Tolerance, avoidance, and escape are known as the three major mechanisms that support the plant to grow in a heat-stress environment. Major heat tolerance mechanisms in wheat are known as stay green, heat shock proteins, and antioxidant defense [42]. Protein synthesis and folding were observed to be interrupted during heat stress. Heat stress also resulted in the production of several stress agents badly affecting transcription, translation, and DNA replication in plants [43]. Plants speed up the production of heat shock proteins as a defense mechanism [44]. Higher activity of antioxidants, such as peroxidases, catalase, and superoxide dismutase, was observed under heat stress. Wheat cultivar showing greater tolerance to heat stress was observed to have higher activity of catalase, ascorbate peroxidase, and S-transferase [45].

Salt stress greatly affects the growth of wheat plants. Salinity stress has a higher impact on the morphology and physiology of wheat plants. Plants having less tolerance to salinity are not suitable for cropping. Potassium transporter (*HKT*) genes have a greater role in achieving salinity tolerance in wheat. Sodium (Na^+) exclusion through *HKT* genes is a major mechanism in wheat to have a salinity tolerance.

OsMYBSs and *AtAB14* are the transcription factors having a role in regulating *HKT* genes, which are considered as the candidate targets for increasing salinity tolerance in wheat [46]. Wheat transformed with a mutated transcription factor, *HaHB4* showed higher water-use efficiency and was more yielding under drought stress [26]. Transgenic wheat expressing *GmDREB1* gene from soybean was also observed to have higher drought tolerance under water-stress conditions [47]. *DREB1A* gene from *Arabidopsis thaliana* was introduced to bread wheat and increased tolerance against water stress in the transgenic wheat was observed. Bread wheat under drought stress was observed to have a higher level of WRKY proteins [48]. Higher expression of *AtHDG11* gene in transgenic wheat resulted in increased water-stress tolerance during drought-stress conditions. Enhanced *TaNAC69* expression in root and leaf of wheat during drought stress was observed [49]. Researchers are working to develop transgenic wheat having various traits/phenotypes by using advanced approaches of biotechnology for the last several decades (**Table 1**). Numbers of transgenic wheat cultivars are being grown in the fields and several more are under trial.

5. CRISPR/Cas9 system in wheat

Gliadins and glutenins are known as the gluten proteins and ingestion of these proteins from barley, rye, and wheat could cause the disease called coeliac disease in humans. The only remedy is to develop gluten-free food. Transgenic wheat which retains baking quality and is safe for coeliac could not be produced using conventional methods because of the complexity of the wheat genome. Coeliac disease (CD) is activated by the immunogenic isotopes mainly gliadins. Gliadin families were downregulated by the use of RNA interference. CRISPR/Cas9 is a targeted gene manipulation tool considered to have a potential role in genetic modification (**Table 2**, [60, 61]). CRISPR/Cas9 system was recently used for gene editing of gliadins. Offsprings with deleted, edited, or silenced gliadins were produced by CRISPR/Cas9. They helped to decrease the exposure of the patient to the CD epitopes [62]. This technology has been used to develop wheat cultivars having gluten genes with inactivated CD epitopes [62, 63].

S. No.	Gene Name	Trait/Phenotype	Reference
1.	*TaMLO*	Powdery mildew resistance	[50]
2.	*TaPHO2-A1*	Improved Phosphorus uptake	[51]
3.	*TaGASR7*	Improved yield	[52]
4.	*TaEDR1*	Powdery mildew resistance	[53]
5.	*TaGW2*	Improved yield	[54]
6.	*TaMs1*	Male sterility	[55]
7.	*TaSBEIIa*	High amylase contents	[56]
8.	*TaLOX2*	Improved quality	[57]
9.	*TaALS*	Herbicide tolerance	[58]
10.	*TaACC*	Herbicide tolerance	[59]

Table 2.
Genome edited wheat developed by CRISPR/Cas9 system.

CRISPR/Cas9 system and TALENS (transcription activator-like effector nuclease) were used in the bread wheat to generate the mutations in three homoeoalleles that encode MLO locus proteins against mildew. Mutations in all three TaMLO were generated by using TALENS which resulted in resistance against powdery mildew. The MLO homoeoalleles (*TaMLOA1*, *TaMLOB1*, and *TaMLOD1*) of bread wheat contributed to the mildew infection. Mutation of MLO alleles resulted in powdery mildew tolerance in wheat [50]. Genome editing was reported in which *pds* (phytoene desaturase) and *inox* (inositol oxygenase) genes in the cell suspension-culture of wheat were targeted. It was demonstrated that the genome-editing technique could also be applied in the cell suspension of wheat [64]. Very recently, various research groups are involved to develop transgenic wheat by using genome-editing technology. Some of the experiments are listed in **Table 2**.

6. Wheat computational analysis

A comprehensive resource for wheat reference genome was developed by International Wheat Genome Sequencing Consortium. The URGI portal (https://wheat-urgi.versailles.inra.fr/) was developed for the breeders and researchers to access the genome sequence data of bread-wheat. InterMine tools, genome browser, and BLAST were established for the exploration of genome sequences together with the additional linked datasets, including gene expression, physical maps, and sequence variation. Portal provided the higher browser and search features that facilitated the use of the latest genomic resources required for the upgradation of wheat [65].

DNA binding with one finger (Dof) transcription factors is known to have an important role in abiotic stress tolerance as well as the growth of plants. Ninety-six TaDof members of the gene family have been studied using computational approaches. By qPCR analysis, it was revealed that TaDof genes were upregulated under heavy metal and heat stress in wheat. Consequently, it could be concluded that detection of amino acid sites, genome-wide analysis, and identification of the Dof transcription factor family could provide us the new insight into the function, structure, and evolution of the Dof gene family [66].

Acknowledgements

This work was supported by funds from the Higher Education Commission of Pakistan.

Author details

Nazia Nahid[1], Parwsha Zaib[1], Tayyaba Shaheen[1*], Kanval Shaukat[2], Akmaral U. Issayeva[3] and Mahmood-ur-Rahman Ansari[1*]

1 Department of Bioinformatics and Biotechnology, GC University–Faisalabad, Faisalabad, Pakistan

2 Department of Botany, University of Balochistan, Quetta, Pakistan

3 M. Auezov South Kazakhstan University, Shymkent, Kazakhstan

*Address all correspondence to: tayaba_pgmb@yahoo.com and mahmoodansari@gcuf.edu.pk

References

[1] FAO. World Food and Agriculture-FAO Statistical Pocketbook. Rome, Italy: FAO; 2015

[2] Akter N, Islam MR. Heat stress effects and management in wheat: A review. Agronomy for Sustainable Development. 2017;**37**(5):1-17

[3] Chaves MS, Martinelli JA, Wesp-Guterres C, Graichen FAS, Brammer SP, Scagliusi SM, et al. The importance for food security of maintaining rust resistance in wheat. Food Security. 2013;**5**(2):157-176

[4] Sharma D, Singh R, Tiwari R, Kumar R, Gupta VK. Wheat responses and tolerance to terminal heat stress: A review. In: Hasanuzzaman M, Nahar K, Hossain M, editors. Wheat Production in Changing Environments. Singapore: Springer; 2019. pp. 149-173

[5] Giraldo P, Benavente E, Manzano-Agugliaro F, Gimenez E. Worldwide research trends on wheat and barley: A bibliometric comparative analysis. Agronomie. 2019;**9**(7):352

[6] Iqbal M, Raja NI, Yasmeen F, Hussain M, Ejaz M, Shah MA. Impacts of heat stress on wheat: A critical review. Advances in Crop Science and Technology. 2017;**5**(1):01-09

[7] Rahaie M, Xue GP, Schenk PM. The role of transcription factors in wheat under different abiotic stresses. Abiotic Stress - Plant Responses and Applications in Agriculture. 2013;**2**:367-385

[8] Lesk C, Rowhani P, Ramankutty N. Influence of extreme weather disasters on global crop production. Nature. 2016;**529**(7584):84-87

[9] Liu B, Asseng S, Müller C, Ewert F, Elliott J, Lobell DB, et al. Similar estimates of temperature impacts on global wheat yield by three independent methods. Nature Climate Change. 2016;**6**(12):1130-1136

[10] Tester M, Langridge P. Breeding technologies to increase crop production in a changing world. Science. 2010;**327**(5967):818-822

[11] He Z, Joshi AK, Zhang W. Climate vulnerabilities and wheat production. In: Pielke RA, editor. Climate Vulnerability: Understanding and Addressing Threats to Essential Resources. Waltham: Academic Press; 2013. pp. 57-67

[12] Sudan J, Sharma D, Mustafiz A, Kumari S. Signaling peptides: Hidden molecular messengers of abiotic stress perception and response in plants. In: Zargar S, Zargar M, editors. Abiotic Stress-Mediated Sensing and Signaling in Plants: An Omics Perspective. Singapore: Springer; 2018. pp. 95-125

[13] Yu TF, Xu ZS, Guo JK, Wang YX, Abernathy B, Fu JD, et al. Improved drought tolerance in wheat plants overexpressing a synthetic bacterial cold shock protein gene SeCspA. Scientific Reports. 2017;**7**(1):1-4

[14] Peña PA, Quach T, Sato S, Ge Z, Nersesian N, Changa T, et al. Expression of the maize Dof1 transcription factor in wheat and sorghum. Frontiers in Plant Science. 2017;**8**:434

[15] Qu B, He X, Wang J, Zhao Y, Teng W, Shao A, et al. A wheat CCAAT box-binding transcription factor increases the grain yield of wheat with less fertilizer input. Plant Physiology. 2015;**167**(2):411-423

[16] Yadav D, Shavrukov Y, Bazanova N, Chirkova L, Borisjuk N, Kovalchuk N, et al. Constitutive overexpression of the TaNF-YB4 gene in transgenic wheat significantly improves grain yield. Journal of Experimental Botany. 2015;**66**(21):6635-6650

[17] He X, Qu B, Li W, Zhao X, Teng W, Ma W, et al. The nitrate-inducible NAC transcription factor TaNAC2-5A controls nitrate response and increases wheat yield. Plant Physiology. 2015;**169**(3):1991-2005

[18] Tian B, Talukder SK, Fu J, Fritz AK, Trick HN. Expression of a rice soluble starch synthase gene in transgenic wheat improves the grain yield under heat stress conditions. In Vitro Cellular & Developmental Biology. Plant. 2018;**54**(3):216-227

[19] Hu M, Zhao X, Liu Q, Hong X, Zhang W, Zhang Y, et al. Transgenic expression of plastidic glutamine synthetase increases nitrogen uptake and yield in wheat. Plant Biotechnology Journal. 2018;**16**(11):1858-1867

[20] Smidansky ED, Meyer FD, Blakeslee B, Weglarz TE, Greene TW, Giroux MJ. Expression of a modified ADP-glucose pyrophosphorylase large subunit in wheat seeds stimulates photosynthesis and carbon metabolism. Planta. 2007;**225**(4):965-976

[21] Zhao XQ, Nie XL, Xiao XG. Over-expression of a tobacco nitrate reductase gene in wheat (Triticum aestivum L.) increases seed protein content and weight without augmenting nitrogen supplying. PLoS One. 2013;**8**(9):e74678

[22] Connorton JM, Jones ER, Rodríguez-Ramiro I, Fairweather-Tait S, Uauy C, Balk J. Wheat vacuolar iron transporter TaVIT2 transports Fe and Mn and is effective for biofortification. Plant Physiology. 2017;**174**(4):2434-2444

[23] Yang Y, Luang S, Harris J, Riboni M, Li Y, Bazanova N, et al. Overexpression of the class I homeodomain transcription factor Ta HDZ ipI-5 increases drought and frost tolerance in transgenic wheat. Plant Biotechnology Journal. 2018;**16**(6):1227-1240

[24] Saint Pierre C, Crossa JL, Bonnett D, Yamaguchi-Shinozaki K, Reynolds MP. Phenotyping transgenic wheat for drought resistance. Journal of Experimental Botany. 2012;**63**(5): 1799-1808

[25] Castiglioni P, Warner D, Bensen RJ, Anstrom DC, Harrison J, Stoecker M, et al. Bacterial RNA chaperones confer abiotic stress tolerance in plants and improved grain yield in maize under water-limited conditions. Plant Physiology. 2008;**147**(2):446-455

[26] González FG, Capella M, Ribichich KF, Curín F, Giacomelli JI, Ayala F, et al. Field-grown transgenic wheat expressing the sunflower gene HaHB4 significantly outyields the wild type. Journal of Experimental Botany. 2019;**70**(5):1669-1681

[27] Shavrukov Y, Baho M, Lopato S, Langridge P. The TaDREB3 transgene transferred by conventional crossings to different genetic backgrounds of bread wheat improves drought tolerance. Plant Biotechnology Journal. 2016;**14**:313-322

[28] García-Molina MD, Giménez MJ, Sánchez-León S, Barro F. Gluten free wheat: Are we there? Nutrients. 2019;**11**(3):487

[29] Bi H, Shi J, Kovalchuk N, Luang S, Bazanova N, Chirkova L, et al. Overexpression of the TaSHN1 transcription factor in bread wheat leads to leaf surface modifications, improved drought tolerance, and no yield penalty under controlled growth conditions. Plant, Cell and Environment. 2018;**41**(11):2549-2566

[30] Godfray HC, Beddington JR, Crute IR, Haddad L, Lawrence D, Muir JF, et al. Food security: The challenge of feeding 9 billion people. Science. 2010;**327**(5967):812-818

[31] Savary S, Willocquet L, Pethybridge SJ, Esker P, McRoberts N, Nelson A. The global burden of

pathogens and pests on major food crops. Nature Ecology & Evolution. 2019;**3**(3):430-439

[32] Menardo F, Praz CR, Wyder S, Ben-David R, Bourras S, Matsumae H, et al. Hybridization of powdery mildew strains gives rise to pathogens on novel agricultural crop species. Nature Genetics. 2016;**48**(2):201-205

[33] Zhu Y, Li Y, Fei F, Wang Z, Wang W, Cao A, et al. E3 ubiquitin ligase gene CMPG 1–V from Haynaldia villosa L. contributes to powdery mildew resistance in common wheat (Triticum aestivum L.). The Plant Journal. 2015;**84**(1):154-168

[34] Liu J, Sun L, Chen Y, Wei L, Hao Y, Yu Z, et al. The regulatory network of CMPG1-V in wheat–Blumeria graminis f. sp. tritici interaction revealed by temporal profiling using RNA-Seq. International Journal of Molecular Sciences. 2020;**21**(17):5967

[35] Eissa HF, Hassanien SE, Ramadan AM, El-Shamy MM, Saleh OM, Shokry AM, et al. Developing transgenic wheat to encounter rusts and powdery mildew by overexpressing barley chi26 gene for fungal resistance. Plant Methods. 2017;**13**(1):1-3

[36] Zhi P, Kong L, Liu J, Zhang X, Wang X, Li H, et al. Histone deacetylase TaHDT701 functions in TaHDA6-TaHOS15 complex to regulate wheat defense responses to Blumeria graminis f. sp. tritici. International Journal of Molecular Sciences. 2020;**21**(7):2640

[37] Shin S, Mackintosh CA, Lewis J, Heinen SJ, Radmer L, Dill-Macky R, et al. Transgenic wheat expressing a barley class II chitinase gene has enhanced resistance against Fusarium graminearum. Journal of Experimental Botany. 2008;**59**(9):2371-2378

[38] Zhang H, Zhang X, Liu J, Niu Y, Chen Y, Hao Y, et al. Characterization of the heavy-metal-associated Isoprenylated plant protein (HIPP) gene family from Triticeae species. International Journal of Molecular Sciences. 2020;**21**(17):6191

[39] Hura T. Wheat and barley: Acclimatization to abiotic and biotic stress. International Journal of Molecular Sciences. 2020;**21**(19):7423

[40] Foulkes MJ, Scott RK, Sylvester-Bradley R. The ability of wheat cultivars to withstand drought in UK conditions: Formation of grain yield. The Journal of Agricultural Science. 2002;**138**(2):153-169

[41] Weldearegay DF, Yan F, Jiang D, Liu F. Independent and combined effects of soil warming and drought stress during anthesis on seed set and grain yield in two spring wheat varieties. Journal of Agronomy and Crop Science. 2012;**198**(4):245-253

[42] Poudel PB, Poudel MR. Heat stress effects and tolerance in wheat: A review. Journal of Biology and Today's World. 2020;**9**(3):1-6

[43] Biamonti G, Caceres JF. Cellular stress and RNA splicing. Trends in Biochemical Sciences. 2009;**34**(3):1 46-153

[44] Gupta SC, Sharma A, Mishra M, Mishra RK, Chowdhuri DK. Heat shock proteins in toxicology: How close and how far? Life Sciences. 2010;**86** (11-12):377-384

[45] Balla K, Bencze S, Janda T, Veisz O. Analysis of heat stress tolerance in winter wheat. Acta Agronomica Hungarica. 2009;**57**(4):437-444

[46] Kunika BK, Singh PK, Rani V, Pandey GC. Salinity tolerance in wheat: An overview. International Journal of Chemical Studies. 2019;**6**:815-820

[47] Shiqing GA, Huijun XU, Xianguo C, Ming C, Zhaoshi XU, Liancheng L, et al.

Improvement of wheat drought and salt tolerance by expression of a stress-inducible transcription factor GmDREB of soybean (Glycine max). Chinese Science Bulletin. 2005;**50**:2714-2723, 2723

[48] Okay S, Derelli E, Unver T. Transcriptome-wide identification of bread wheat WRKY transcription factors in response to drought stress. Molecular Genetics and Genomics. 2014;**289**(5):765-781

[49] Xue GP, Bower NI, McIntyre CL, Riding GA, Kazan K, Shorter R. TaNAC69 from the NAC superfamily of transcription factors is up-regulated by abiotic stresses in wheat and recognises two consensus DNA-binding sequences. Functional Plant Biology. 2006;**33**(1):43-57

[50] Wang Y, Cheng X, Shan Q, Zhang Y, Liu J, Gao C, et al. Simultaneous editing of three homoeoalleles in hexaploid bread wheat confers heritable resistance to powdery mildew. Nature Biotechnology. 2014;**32**(9):947-951

[51] Ouyang X, Hong X, Zhao X, Zhang W, He X, Ma W, et al. Knock out of the PHOSPHATE 2 gene TaPHO2-A1 improves phosphorus uptake and grain yield under low phosphorus conditions in common wheat. Scientific Reports. 2016;**6**:29850

[52] Zhang Y, Liang Z, Zong Y, Wang Y, Liu J, Chen K, et al. Efficient and transgene-free genome editing in wheat through transient expression of CRISPR/Cas9 DNA or RNA. Nature Communications. 2016;**7**:12617

[53] Zhang Y, Bai Y, Wu G, Zou S, Chen Y, Gao C, et al. Simultaneous modification of three homoeologs of TaEDR1 by genome editing enhances powdery mildew resistance in wheat. The Plant Journal. 2017;**91**:714-724

[54] Wang W, Mauleon R, Hu Z, Chebotarov D, Tai S, Wu Z, et al. Genomic variation in 3,010 diverse accessions of Asian cultivated rice. Nature. 2018;**557**(7703):43-49

[55] Okada A, Arndell T, Borisjuk N, Sharma N, Watson-Haigh NS, Tucker EJ, et al. CRISPR/Cas9-mediated knockout of Ms1 enables the rapid generation of male-sterile hexaploid wheat lines for use in hybrid seed production. Plant Biotechnology Journal. 2019;**17**:1905-1913

[56] Li J, Jiao G, Sun Y, Chen J, Zhong Y, Yan L, et al. Modification of starch composition, structure and properties through editing of TaSBEIIa in both winter and spring wheat varieties by CRISPR/Cas9. Plant Biotechnology Journal. 2020;**19**:937-951

[57] Zong Y, Wang Y, Li C, Zhang R, Chen K, Ran Y, et al. Precise base editing in rice, wheat and maize with a Cas9-cytidine deaminase fusion. Nature Biotechnology. 2017;**35**:438-440

[58] Zhang R, Liu J, Chai Z, Chen S, Bai Y, Zong Y, et al. Generation of herbicide tolerance traits and a new selectable marker in wheat using base editing. Nature Plants. 2019;**5**:480-485

[59] Zhang Y, Malzahn AA, Sretenovic S, Qi Y. The emerging and uncultivated potential of CRISPR technology in plant science. Nature Plants. 2019;**5**:778-794

[60] Schaart JG, van de Wiel CC, Lotz LA, Smulders MJ. Opportunities for products of new plant breeding techniques. Trends in Plant Science. 2016;**21**(5):438-449

[61] Van de Wiel CC, Schaart JG, Lotz LA, Smulders MJ. New traits in crops produced by genome editing techniques based on deletions. Plant Biotechnology Reports. 2017;**11**(1):1-8

[62] Jouanin A, Gilissen LJ, Schaart JG, Leigh FJ, Cockram J, Wallington EJ, et al. CRISPR/Cas9 gene editing of

gluten in wheat to reduce gluten content and exposure—Reviewing methods to screen for coeliac safety. Frontiers in Nutrition. 2020;**7**:51

[63] Jouanin A, Schaart JG, Boyd LA, Cockram J, Leigh FJ, Bates R, et al. Outlook for coeliac disease patients: Towards bread wheat with hypoimmunogenic gluten by gene editing of α-and γ-gliadin gene families. BMC Plant Biology. 2019;**19**(1):1-6

[64] Upadhyay SK, Kumar J, Alok A, Tuli R. RNA-guided genome editing for target gene mutations in wheat. G3: Genes, Genomes, Genetics. 2013;**3**(12):2233-2238

[65] Alaux M, Rogers J, Letellier T, Flores R, Alfama F, Pommier C, et al. Linking the international wheat genome sequencing consortium bread wheat reference genome sequence to wheat genetic and phenomic data. Genome Biology. 2018;**19**(1):1

[66] Liu Y, Liu N, Deng X, Liu D, Li M, Cui D, et al. Genome-wide analysis of wheat DNA-binding with one finger (Dof) transcription factor genes: Evolutionary characteristics and diverse abiotic stress responses. BMC Genomics. 2020;**21**(1):1-8

Section 2

Biotic Stress in Wheat

Chapter 2

Breeding Wheat for Biotic Stress Resistance: Achievements, Challenges and Prospects

Harmeet Singh Bakala, Kamalpreet Singh Mandahal, Ankita, Loveleen Kaur Sarao and Puja Srivastava

Abstract

Wheat (*T. aestivum*) is one of the key food grain crops and is a prominent source of calories and proteins globally. In addition to mushrooming population and rising abiotic stresses in this ongoing climate change era, biotic stresses pose a great threat to wheat production over the globe. Fungal diseases such as rusts, mildew, along with pests like aphid, hinder the potential yield performance of the elite wheat cultivars to a huge extent. The complex nature of plant-parasite interactions is shown to be the decisive factor for the ultimate resistance expression in wheat. However, the advancement of molecular genetics and biotechnology enabled the replacement of the tedious, time and resource consuming cytogenetic analyses of locating APR and ASR genes using molecular mapping techniques. Continuous efforts have been made to mine resistance genes from diverse genetic resources such as wild relatives for combating these diseases and pests, which are repositories of R genes. Additionally, they offer a promising source of genetic variation to be introgressed and exploited for imparting biotic stress tolerance in cultivated wheat. Though just a handful of R-genes are cloned and molecularly characterized in wheat so far, more than 350 resistance genes for various diseases have been identified and successfully introgressed into elite varieties around the globe. Modern genomics and phenomic approaches coupled with next-generation sequencing techniques have facilitated the fine-mapping as well as marker aided selection of resistance genes for biotic stress resistance wheat breeding.

Keywords: Biotic stress, Durable resistance, Genomics, R-genes, Wheat rusts, Wild relatives

1. Introduction

Wheat is one of the most important cereal staple food crops in the world, both in terms of food production and for providing the total amount of food calories and protein in the human diet [1]. It is believed that bread wheat originated in south western Asia from where it spread to other regions of Asia, Europe, Africa and America [2]. Wheat has adapted itself to diverse climatic conditions and, as such, is grown over a range of altitudes and latitudes under irrigated, severe drought and wet conditions. The global demand for wheat is projected to rise by 60% by 2050 because of the increase in the world's human population and changing livelihoods.

Wheat production has been threatened by unexpected abiotic and biotic stresses due to abrupt environmental changes or movement of pathogens. The monoculture of modern wheat cultivars with low genetic diversity has resulted in pathogen resurgences, which threaten wheat supplies [3].

Biotic stress in plants is caused by several living organisms namely fungi, virus, insects, nematodes, arachnids and weeds. Unlike the stresses caused by environmental factors i.e. abiotic stresses (heat and drought), the biotic stress agents directly affect the host growth and development by depriving them of nutrition resulting into reduced plant vigor and in extreme cases, even death of the host. From the agricultural context, biotic stress has major contribution in pre as well postharvest losses. Of the nearly 200 diseases and pests that have been documented, 50 are considered economically important because of their potential to damage crops and affect farmers' incomes [4]. Among biotic stresses, pathogenic fungi represent a significant challenge to wheat production globally. The major diseases in wheat involves stripe rust, stem rust, leaf rust, powdery mildew, head blight etc. Historically, yellow rust has caused and is presently causing significant and severe losses in susceptible wheat cultivars worldwide [5]. The major insect-pests attacking wheat are aphid, hessian fly, green bug and borers etc.

In this chapter, the major diseases and pests detrimental to wheat crop along with the molecular basis of stress resistance will be discussed. Moreover, the remarkable global milestones being achieved along with some important tools and prospects for mitigating with these economically important diseases and pests will be focused.

2. Biotic stress resistance in wheat

2.1 Types of disease resistance

There are basically two types of genetic resistances as described by Vander Plank [6] for the different diseases in wheat i.e. Qualitative/Vertical resistance and Quantitative/Horizontal resistance.

2.1.1 Qualitative (vertical) resistance

It is specified to pathogen races controlled by a single or few genes i.e. monogenic or oligogenic. Race-specific is used to describe resistance that interacts differentially with different pathogenic races i.e. it is applied both to complete resistance and components of incomplete resistance that so interact [7]. This kind of resistance is easily detectable with specific pathogenic races or pathotypes which are controlled by genes with major effects. In wheat rust pathosystems, these resistances are recognized by characteristic low infection types. Most of these genes can be detected in seedling evaluations using specific pathotypes. For every resistance gene in the host plant, there is a corresponding virulent gene in the pathogen as stated by gene for gene hypothesis. However the ability of a virulent gene to mutate to avirulent gene, no longer recognizable by the corresponding resistance gene, implies a type of resistance termed race-specific resistance.

2.1.2 Quantitative (horizontal) resistance

This kind of resistance varies in continuous way among the different phenotypes of the host population, ranging from almost imperceptible to quite strong resistance response. The resistance expression depends upon the genotype and environment, where pathogen is the part of that environment. The environment can considerably affect its durability also [7]. Partial resistance is supposed to be under polygenic control and such resistance will be race-nonspecific. Being controlled by minor genes, the quantitative resistance has complex genetic basis which operates against all the pathotypes/races of that specific pathogen. Race-nonspecific resistance is mainly effective at the post-seedling and adult plant stages and adult plant resistance (APR) is often detected as field resistance [8]. The best known APR genes in wheat are *Sr2* (stem rust resistance gene) and *Lr34*, a gene that provides resistance to leaf and stripe rust and powdery mildew. These genes have been used in commercial wheat varieties for almost 100 years. *Sr2* and *Lr34* have provided partial resistance for decades over large areas and under prolonged disease pressure in the field, proving their durability. Adding to complexity, Ug99 had a very wide spectrum of virulence towards most of the commonly used R genes and rapidly evolved virulence to the important R genes (*Sr24* and *Sr36*) which has impeded the initial emergency breeding response to incorporate resistance to this strain [9].

2.2 Types of insect resistance

Insect resistance on the other hand is typically governed by three main mechanisms.

2.2.1 Single or oligogenic resistance

Single or oligogenic resistance has been observed against some insects such as Hessian fly in wheat. Such resistance is governed by a single or few major genes. This type of resistance has also been reported against Russian wheat aphid and green bug.

2.2.2 Polygenic resistance

Several genes with small additive effects govern the resistant response against some insects. The resistance observed against cereal leaf beetle in wheat is of this type.

2.2.3 Cytoplasmic resistance

Cytoplasmic resistance against insects has not been reported in case of wheat. However, in maize resistance against European corn borer is governed by cytoplasmic genes. Another case of cytoplasmic resistance is observed in lettuce against root aphid.

3. Major diseases of wheat

There are many diseases found in wheat caused by different microorganisms from fungi to bacteria and viruses. But only a few of them caused by pathogenic fungi are economically important with global implications. The major diseases in wheat (**Table 1**) are stripe rust, leaf rust, stem rust, powdery mildew, loose smut, Fusarium head blight (FHB) and more recently wheat blast (WB) also. Besides Stem rust, which is under control to some extent, Leaf rust and yellow rust have the potential to affect production levels up to 60 and 43 million hectares respectively in Asia if susceptible cultivars were grown [10]. Though fungicidal applications offer control, their use is an added cost to farmers besides being unsafe environmentally. Hence growing resistant cultivars is the most effective and efficient control strategy

S no	Disease	Causal Pathogen	Behavior	No. of R-genes identified
1.	Stripe rust (yellow)	*Puccinia striiformis* f. sp. *tritici*	Biotrophic	95
2.	Leaf rust (brown)	*Puccinia triticina*	Biotrophic	80
3.	Stem rust (black)	*Puccinia graminis* f. sp. *tritici*	Biotrophic	67
4.	Powdery mildew (PM)	*Blumeria graminis* f. sp. *tritici*	Biotrophic	70
5.	Karnal Bunt (KB)	*Tilletia indica*	Biotrophic	6
6.	Fusarium head blight (FHB)	*Fusarium graminearum*	Necrotrophic	7
7.	Wheat blast (WB)	*Magnaporthe oryzae* pathotype *triticum*	Necrotrophic	5
8.	Loose Smut (LS)	*Ustilago tritici*	Biotrophic	10

Table 1.
Major diseases of wheat with their respective behavior and number of resistance genes identified for each disease (up to 2020).

[11]. The rusts and mildew diseases are caused by biotrophic fungi (survive by obtaining nutrients from living plant tissues). Among these, *Puccinia* rusts continue to affect and threaten the world's wheat production [12], although powdery mildew has also emerged as an economically important disease. In case of stem rust, the emergence of *Ug99* group of stem rust races placed it among one of the most significant threats to global wheat production [13].

The other diseases like FHB and WB are caused by necrotrophic fungi (facultative parasites feeding on dead tissue during unavailability of living plants). Wheat blast was first identified in Parana, Brazil in 1985 [14]. It is also of utmost significance as WB outbreaks in Bangladesh [15] and more recently in Africa [16] have attracted immediate global attention from the wheat scientists.

Another economically significant disease, Karnal Bunt (KB) of wheat was first reported in Karnal, India [17], soon extended to Northern and Central India. Later, KB was found to occur in Nepal, Pakistan, Iraq, Afghanistan, South Africa, Mexico and USA [18]. The pathogen is seed, soil and airborne in nature, therefore difficult to control after it is introduced and then established over a region. Although host plant resistance is the most effective and economic method of its management but development of KB resistance varieties is difficult task owing to limited genetic variability in hexaploid wheat [19], quantitative inheritance and considerable impact of environment on KB resistance screening [20].

4. Major insect-pests of wheat

Various insect pests delimit the yields of wheat crop in different agro-climatic zones. Some of these insect pests are foliar aphid complex in irrigated wheat, root aphids in loose soils, pink stem borers in fields having rice stubbles, cut worms in residues, termites in raised beds and brown mites in rainfed conditions [21]. Six different species of aphids are reported to attack cereals. Out of these, Russian wheat aphid and bird cherry-oat aphid are important pests of wheat. The Russian wheat aphid (*Diuraphic noxia*) is a sucking pest of wheat. Aphid attack is characterized by leaf rolling which is the result of toxic injection by the aphid. The rolled leaves serve as a protection site for the insects. Yield losses up to 40% have been reported in case of aphid infection [22]. The bird cherry oat aphid (*Rhopalosiphum padi*) has been reported to affect wheats all over the world. Feeding symptoms are almost absent. Yield losses due to *R. padi* dependent upon the crop stage at which insect attacks. High yield losses upto 24–65% have been reported in case the attack occurs at seedling stage. Losses decrease if attack occurs at later stages [23]. The aphid is also reported to cause significant indirect losses as it is a vector of Barley Yellow Dwarf Virus (BYDV), which is the most important viral disease in cereals. Greenbug (*Schizaphis graminum*) is another sucking pest of the wheat aphid complex. The green bug feeds on wheat leaves and stems, extracting sap from the phloem. Injection of toxins concomitant with feeding further reduces the chlorophyll content thereby inversely affecting the carbon assimilation and overall plant development [23–25].

Cephus spp., the wheat stem sawfly has also been reported to cause major losses in wheat. The adult females oviposit into the young stems of wheat. Upon hatching within the stem, the larvae feed voraciously moving up and down in the stem. When the plant attains maturity, larvae migrate to the basal portion of the stem and build a hibernaculum. The stem above the hibernaculum weakens and breaks [26]. The Hessian fly (*Mayetiola destructor*) is another major pest of wheat crop. Larvae damage stems of plants, thereby preventing internode elongation and disrupting nutrient transport. Significant losses (upto 40%) have been reported upon sawfly attack [27].

5. Molecular basis of disease resistance in wheat

Wheat is an allopolyploid, means a polyploid species that resulted from interspecific or intergeneric hybridization of two or more genomes from different species. Polyploidy, a common form of plant evolution, is associated with promoting the genetic diversity that facilitates adaptation to a range of environments. Because wheat is a global crop, it is under continuous exposure to a large variety of parasite species and strains, many of which have the ability to move around the globe. Long-term co-evolution between plants and their pathogens has equipped plants with a sophisticated multi-layered immune system to guard themselves against pest and pathogens [28]. Specificity between pathogenic variants (races) and plant genotypes (cultivars) follows gene for gene-for-gene interactions, whose outcome is conditioned by alleles of a gene regulating resistance (*R* gene) in plant and alleles of its corresponding gene regulating avirulence (*Avr* gene) in pathogen [29]. The plant immune system is typically described in terms of two components: pattern triggered immunity (PTI) which is activated by recognition of microbial or pathogen-associated molecular patterns (MAMPs or PAMPs) and effector-triggered immunity (ETI) involving gene for gene kind of resistance [30, 31]. ETI is often based on the recognition of cytosolic effectors by immune receptors with a conserved nucleotide-binding domain (NBARC) and a leucine-rich repeat domain (LRR) also called NLRs. This type of resistance is usually associated with a hypersensitive response (HR) localized to infection sites. To date, only a handful of these biotic stress resistance genes have been isolated and cloned in wheat (*T. aestivum*). Donors of the *R* genes are genetically diverse, including species in the primary gene pool (*Triticum* spp.), secondary gene pool (e.g. *T. timopheevii*), and tertiary gene pool (e.g. *Aegilops*, *Secale*, and *Thinopyrum*).

5.1 NBS: LRR proteins - basis of race-specific/seedling/all stage resistance (ASR)

A few resistance genes have been cloned for race-specific resistance in wheat so far, which belong to a conserved gene family encoding NBS-LRR (Nucleotide binding site-leucine-rich repeat) proteins, also known as R-proteins (NLR) [30]. For example, powdery mildew genes, *Pm3* and *Pm8* and leaf rust resistance genes *Lr10* and *Lr21*. These R-proteins impart complete but race specific resistance. NBS-LRR proteins are a conserved class of immune receptors that directly or indirectly recognize pathogen-specific effector proteins. These proteins are secreted by pathogens into the host cell to suppress defense response and to establish infection. Recognition of effectors by NBS–LRR proteins triggers a signaling cascade resulting in a strong resistance response called hypersensitive reaction (HR) [31]. HR eventually leads to death of the infected host cell by this means preventing further spread of the pathogen [32]. Since this type of disease resistance depends on the recognition of specific pathogen effectors, even point mutations within effector genes or their loss can disrupt recognition by the corresponding NBS–LRR protein. Such mutations in pathogen effectors result in the emergence of new virulent pathogen races and breakdown of disease resistance. Mutated pathogen spores that avoid recognition by the corresponding R gene will have a huge selective advantage facilitating their rapid multiplication. Dispersal of fungal pathogens by wind over long distances adds to the quick spread of newly evolved virulent pathogen strains. Ug99, for instance, spread out from Kenya to South Africa and the Near East in less than a decade.

So far, only 31 genes have been cloned (**Table 2**) for biotic stress resistance (30 for disease resistance) from bread wheat and its wild relatives. Among these, most of the genes impart race specific resistance to the plant. These R-genes encode proteins with an NBS-LRR domain with a coiled-coil (CC) domain. This type of gene typically shows a greater degree of variation in LRR-encoding sequences [60, 61]. This is consistent with the idea that the LRR-encoding sequence is important for target specificity [61, 62]. The sequence variation in NBS-encoding region can also play significant role in specificity. For powdery mildew resistance, *Pm3* locus encodes seven alleles (*Pm3a–Pm3g*) providing resistance to different races of *Blumeria graminis* f. sp. *tritici* [63]. Sequence analysis indicated that the *Pm3* alleles evolved either by gene conversion/recombination or by single point mutations within the NBS and LRR regions [61].

5.2 Transporter proteins: basis of durable/adult plant resistance (APR)

Due to rapid pathogen evolution, R gene resistance is often not durable. One strategy to increase the longevity of disease resistance in wheat cultivars is to pyramid several R genes in one cultivar. To overcome such resistance gene stacks, simultaneous mutations in several effector genes would be required in one single pathogen spore. Race-non-specific resistance is supposed to be more durable when deployed in agriculture. Such kind of resistance mechanism sometimes may also be effective against multiple pathogens. These are normally quantitative traits conferring partial resistance that is able to slow down disease development. For example *Lr34*, *Yr36*, and *Pm21*. *Lr34* confers non-specific, partial, and slow rusting resistance, and has been deployed worldwide, maintaining its effectiveness in agriculture for decades. Due to its role in conferring resistance to pathogens other than leaf rust, it is also known as *Yr18, Pm38, Sr57* and *Bdv1* for resistance to stripe rust, powdery mildew, stem rust, and barley yellow dwarf virus, respectively [64]. The successful cloning of *Lr34, Yr36, and Lr67* revealed these APRs encode an ABC transporter, a kinase-START protein, and a hexose transporter, respectively (**Table 2**). They appear to each have their own resistance mechanism, function

S.no	Gene	Biotic stress	Protein type	Reference
1.	*Lr21*	Leaf rust	NLR	[33]
2.	*Lr10*		NLR	[34]
3.	*Lr1*		NLR	[35]
4.	*Lr34/Yr18/Sr57/Pm38*		ABC[1] transporter	[36]
5	*Lr67/Yr46/Sr55/Pm46*		Hexose transporter	[37]
6	*Lr22a*		NLR	[38]
7	*Yr36/WKS1*	Stripe/yellow rust	Kinase-START[2]	[39]
8	*Yr7*		NLR	[40]
9	*Yr5a*		NLR	[40]
10	*Yr5b*		NLR	[40]
11	*Yr15*		Tendem kinase-pseudokinase	[41]
12	*YrAS2388*		NLR	[42]
13	*Sr33*	Stem rust	NLR	[43]
14	*Sr35*		NLR	[44]
15	*Sr50*		NLR	[45]
16	*Sr22*		NLR	[46]
17	*Sr45*		NLR	[46]
18	*Sr13*		NLR	[47]
19	*Sr21*		NLR	[48]
20	*Sr46*		NLR	[49]
21	*SrTA1662*		NLR	[49]
22	*Sr60/WTK2*		Tendem kinase	[50]
23	*Sr26*		NLR	Zhang et al. (under review)
24	*Sr61*		NLR	Zhang et al. (under review)
25	*Pm3*	Powdery Mildew	NLR	[51]
26	*Pm8*		NLR	[52]
27	*Pm2*		NLR	[53]
28	*Pm21*		serine/threonine protein kinase	[54, 55]
29	*Pm60*		NLR	[56]
30	*WFhb1–1 (Qfhb1)*	Fusarium head blight (FHB)	PFT[3]- chimeric lectin	[57, 58]
31	*H13*	Hessian fly	CC-NB-ARC-LRR	[59]

[1]*ABC- ATP binding cassette.*
[2]*START- Steroidogenic acute regulatory protein-related lipid transfer domain.*
[3]*PFT- Pore-forming toxin.*

Table 2.
List of major cloned resistance genes in wheat for different biotic stresses.

constitutively and often increase the basal level of resistance of the host, which is different from the recognition based NLRs.

6. Insect resistance in wheat

6.1 Resistance categories

Responses which govern insect resistance in plants can be classified into three categories. Tolerance can be defined as the response of plant which allows the plant to survive insect damage with low or no damage to the yield. Tolerance is generally governed by a complex set of genetic traits. Tolerance does not affect the overall survival of insects thereby poses no selection pressure. Tolerance has been reported in a number of crops [65, 66]. The non-preference of a plant by insect pest or antixenosis is another mechanism used by plants against insects. Generally, antixenosis is manifested by some morphological or chemical factors which hinder feeding of the pest and sometimes rejection as host. Antibiosis, the third category, can be defined as the condition when pest health and reproduction are negatively affected by the resistant plant. Most of the resistance observed in field (up to 90%) is due to antibiosis.

6.2 Resistance mechanisms

Over the due course of evolution traits for direct and indirect defense mechanisms against insect attacks have developed in plants. The classification of these mechanisms has been further done as direct mechanisms and indirect mechanisms. Structural barriers constitute the direct defenses. Tissue toughness, glandular and non-glandular trichomes and plant pubescence are included in these types of defenses. Allelochemicals in plant tissues are also included in direct defenses. These exhibit toxic, anti-feedant, and repellent effects on the attacking arthropods. The digestive enzyme inhibitors, cyanogenic glycosides, glucosinolates, lectins, glucosinolates, terpenoids and alkaloids are involved in this [67, 68]. An extensive review of constitutive & induced morphological & chemical plant defenses has been done [65, 66, 69, 70]. These defenses mediate antixenosis & antibiosis. Volatile organic compounds constitute the indirect defenses. The plants which are damaged by pest arthropod release these compounds. These compounds lead to attraction of arthropod predators & parasitoids or the ones that cause repelling of oviposition of pest arthropods [71]. The specific plant indirect defense responses are represented by herbivore associated molecular patterns (HAMPs). These are the responses to the specific herbivore derived elicitors. This occurs in the in oral or ovipositor secretions. These facilitate indirect defenses against herbivores [72]. The widely researched HAMPs are the insect fatty acid plant amino acid conjugates. These are obtained from the lepidopterous larvae [71, 73].

6.3 Constitutive and induced resistance genes

The arthropod selects host plant tissue substrate based on well-coordinated interactions occurring within evolutionarily conserved protein(s) which are encoded by attacking arthropod & responding host plant. The arthropod successfully manipulates the host plant as a result of suitable arthropod-plant interactions. When there is incompatibility in the arthropod-plant interaction, the arthropod does not succeed resulting in the survival of the plants attacked [74]. The plant and fungal endophytic genes are expressed in both the interactions. These are expressed constitutively or via induced defense responses. These occur following herbivory and find involvement in arthropod resistance [75, 76].

Under field conditions, resistance has been explained more clearly by the effects which are controlled by the constitutive genes. This is concluded based on the limited research done till date. The effects owing to the induced gene expression do not contribute much in this [77, 78]. The generation of reactive oxygen species and the signal cascades which involve salicylic acid (SA), jasmonic acid (JA), abscisic acid, ethylene and gibberellic acid occurs in plants as a response to arthropod herbivory. Direct and indirect defense proteins are resulted by the downstream production [79–82]. The aphid bacterial endosymbionts could also lead to defense signals [83]. Jasmonic acid based transcriptomes are elicited by the plant tissue damage caused by arthropods with chewing mouthparts. On the contrary, arthropods with piercing-sucking mouthparts induce the jasmonic acid- salicylic acid-based transcriptomes [71]. Recent documentation has been done of the jasmonic acid- salicylic acid signaling induced by both types of herbivory and jasmonic acid- salicylic acid cross talk [68, 74, 84, 85]. The expression of several plant genes which are produced in the initial responses to arthropod herbivory are controlled by the JA, 12- oxo-phytodienoic acid, and jasmonoyl-amino acid conjugates (which are governed by zinc finger protein expressed in inflorescence meristem) repressor proteins [86]. Several defense allelochemicals are produced by the defense response gene upregulation. This occurs via JA and some other pathways [69]. Scanty information is available regarding the arthropod induced expression of the plant metabolism genes. There are very few evidences indicating the down regulation of few of these genes. This is reported to occur in the beginning just after the arthropod herbivory sets in and later on upregulated in the ensuing days [84, 87].

The identification of arthropod pest elicitors of resistance genes is yet to done. An undefined elicitor protein of *Diuraphis noxia* is recognized by the wheat plant receptors. *D. noxia* is recognized by plant-signaling gene products feeding in incompatible interactions [88]. Secondary metabolites possessing the Hydroxamic acids (Hx) (1,4-benzoxazin-3-ones) group, find involvement in the resistance of certain cereals against bacteria, fungi and several insects including aphids [89]. In the seed, Hydroxamic acids (Hx) are absent. This increases after germination. The young seedlings exhibit the concentration peak [90]. This is basically located in the mesophyll protoplasts, the vascular bundles [91] and in the sieve elements [92]. In the mature plants, the Hx levels decline after the seedling stage. Even then, the young tissue still exhibits a high concentration of Hx [90]. In the plants, the Hx compounds occur as 2-β-O-D-glucopyranosides [90]. When the tissue is injured, these are enzymatically hydrolyzed by endo-β-glucosides to DIMBOA (2,4-dihydroxy-7-methoxy-1,4-benzoxazin-3- one) [92]. DIMBOA is the main Hx aglucone in the wheat extracts. It leads to antibiosis, decreased performance, feeding deterrence and reduced reproduction in aphids [93].

An enhancement in the overall activity of several enzymes was observed. All the enzymes such as superoxide dismutase, phenylalanine ammonia lyase glutathione reductase, and polyphenol oxidase have a major role in the defense of plants towards the feeding of aphid [94]. An early defense strategy is mounted by the Hessian fly-resistant *Ae. tauschii*. The production of anti-feedant proteins (lectins), secondary metabolites and ROS radicals is involved in this strategy. These successfully counter the larval extra oral salivary plant cell degrading proteases, lead to fortification of the cell wall and prevention of the Hessian fly larvae from establishing permanent feeding sites [95].

There are different types of carbohydrate binding proteins known as lectins which are present in tissues of plants. Resistance building potential is possessed by these lectins for wheat against insects. To tackle HF, the identification of genes leading to production of this type of lectins seems a potential method. The genes include Hfr-2 called as HF destructor. This is expressed in the leaf sheaths of the resistance genotypes [96] On similar lines, the mannose binding lectins serve as storage protein and accumulate in the phloem sap. This might act against HF. Anti-insect properties are possessed by these lectins. This is attributed to the accumulation of lectin in the midgut of insects, killing them instantly. Another defensive mechanism present in resistant varieties of wheat is the production of Wci-1 mRNAs and Hfr-1. This occurs in response to the attack of HF larvae. The Hfr-1 gene is known as the defender gene against HF. It has the ability to control crop from severe attack [97]. The identification of arthropod pest elicitors of resistance genes is yet to done. An undefined elicitor protein of *Diuraphis noxia* is recognized by the wheat plant receptors. *D. noxia* is recognized by plant-signaling gene products feeding in incompatible interactions [88]. An Avirulence (*Avr*) gene is there on the parasites side. This encodes one of the several effector proteins that the parasite applies to the plant to help in colonization. A Resistance (R) gene is there on the plant's side. It mediates a surveillance system which detects the Avr protein. The detection is done either directly or indirectly. It triggers effector-triggered plant immunity. The arthropods are responsible for a significant proportion of plant biotic stress but even then they have not been integrated into important models of plant immunity that arise from plant pathology. The absence of molecular evidence for arthropod Avr effectors has been a limiting factor. This evidence was discovered in a plant pathogen around thirty years back. Now, there is evidence for arthropods with the cloning of the Hessian fly's vH13 *Avr* gene. Resistance against RWA is supposed to be induced by gene-for-gene model. The resistant gene produces a protein in this mechanism. This protein contains nucleotide binding site-leucine rich repeat (NBSLRR) domain [98, 99]. Firstly, this NBSLRR domain recognizes and then interacts with cognate Avr protein which is produced by the respective insect [100]. It has been reported that another domain (serine/threonine-protein kinases: STKs) is produced by *Dn* genes. This confers resistance against the RWA [101].

7. Sources of biotic stress resistance in wheat

Wheat belongs to the kingdom Plantae and family Poaceae. It is a long day and a self-pollinated crop. The bread wheat (*Triticum aestivum*) genome is one of the most challenging plant genomes to study. It is highly repetitive (~85%) and approximately 15.4–15.8 Gbp in size, which is five times larger than the human genome [102]. The genus Triticum contains 10 species, out of which six are cultivated and four are wild. Hexaploid wheat (*T. aestivum*) genome ($2n = 6x = 42$) encompasses A, B and D sub-genomes which is advantageous for providing useful genetic diversity for crop improvement. There are three ploidy levels in *Triticum* and *Aegilops* (encompassing cultivated wheats and their progenitors) genera with 2n chromosomes 14, 28, 42 and the basic chromosome $x = 7$ in all the species. Other genera of Poaceae such as Secale, Hordeum, Dasopyrum, Agropyron, Elymus, Leymus, Elytrigia, and Thinopyrum are also important for introgression of useful variability into cultivated wheats. On the basis of their genomic constitution, the wild relatives of wheat can be classified into primary, secondary, and tertiary gene pools [103, 104]. These gene pools are affluent source of genes for disease and pest resistance, mitigating abiotic stresses and micronutrient enrichment in wheat. These three gene pools of wheat as sources of resistance can be described as follows:

1. The primary gene pool consists of species sharing homologous genomes with cultivated wheat. This group includes land races of *T. aestivum*, *T. turgidum* and donor species of the A and D genomes of bread wheat-*T. monococcum, T. urartu, T. boeoticum and Ae. tauschii*. Gene transfer from these species can be achieved by direct hybridization, backcrossing, and selection [104]. Just embryo rescue in certain cases is necessary to produce F_1 hybrid. Many genes conferring resistance to diseases and insect pests have been transferred using this method and several of them are still being exploited in cultivar improvement [105, 106]. Among genetic resources, landraces has been reported a crucial germplasm pool contributing to the genes for grain yield [107, 108] high protein content and tolerance to biotic/abiotic stresses [109]. The green revolution semi-dwarfing genes (*Rht- B1b* and *Rht-D1d*) [110] and other semi-dwarfing gene, *Rht8c*, has been a significant contribution of the landraces. The *Rht* dwarfing gene that was available through the Japanese variety 'Norin10' originating from a Japanese landrace Shiro Daruma [111]. Later, these dwarfing genes were utilized by Dr. Norman E. Borlaug to develop the high-yielding semi-dwarf wheat varieties triggering the Green Revolution in late 1960s.At Punjab Agricultural University (PAU), Ludhiana, India, an active collection of 280 *Ae. tauschii* accessions is being maintained. These accessions have been found to carry resistance genes for various biotic stresses including leaf rust, stripe rust, powdery mildew, and Karnal bunt. Ae. tauschii has a very high level of KB resistance.

2. The secondary gene pool of bread wheat includes the polyploid *Triticum* and *Aegilops* species that have at least one genome in common with wheat. Gene transfer from these species by homologous recombination is possible, if the target gene is located on a homologous chromosome. However, if the genes are present in a non-homologous genome, special cytogenetic manipulations are required. These species have contributed many resistance genes that are being used in cultivar development [103]. At PAU, the genes for disease resistance and HMW glutenin subunits have been successfully transferred from several *Triticum* and *Aegilops* species into wheat and durum cultivars with direct hybridization and backcrossing [112, 113].

3. Species belonging to the tertiary gene pool are more distantly related. Their chromosomes are not homologous to those of wheat. Gene transfer from these species cannot be achieved by homologous recombination, chromosome pairing, and recombination between wheat chromosome and alien chromosomes [103, 104]. Special cytogenetic techniques (in-situ hybridization) are required to ensure compensating transfers with least linkage drag for commercial exploitation of introgressed derivatives. Even though such transfers may include an entire chromosome arm or part of an arm, these have been successfully bred into commercial wheat cultivars because the alien chromosome segment genetically compensates for the missing wheat segment.

8. Major techniques for inducing biotic stress resistance

The route maps followed for a trait improvement particularly stress resistance, both biotic and abiotic remain the same. The **Figure 1** graphically depicts various tools and techniques that can be utilized with efficient and effective manner for tackling different biotic stresses in wheat.

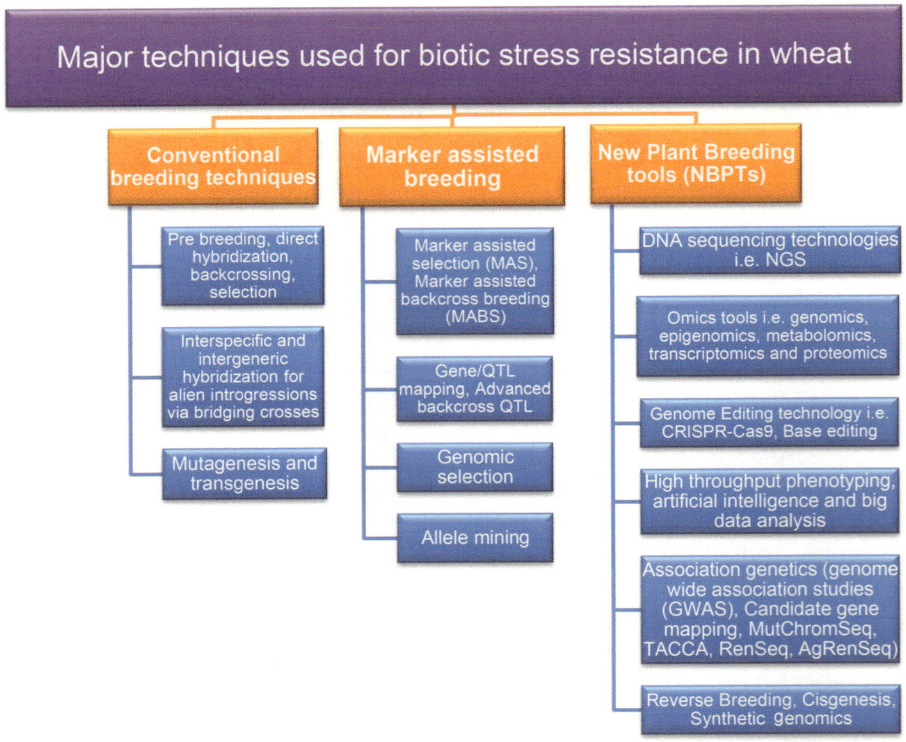

Figure 1.
Some major tools and techniques (both in use and under exploration) in wheat breeding for biotic stress resistance.

9. Present scenario

9.1 Fungal diseases

So far, more than 240 rust resistance genes have been characterized and formally designated in wheat or its relatives; most being race-specific resistance genes. At least 67 of these genes are designated as *Sr* resistance genes [105, 114, 115]. *Sr31* was one of the most widely utilized race-specific *Sr* resistance genes [116]; however, its presence at the International Maize and Wheat Improvement Center (CIMMYT) has been drastically reduced following testing against Ug99 races in Kenya. Evolution of virulence against *Sr31* with the emergence of Ug99 led to stem rust susceptibility in most of the wheats grown around the globe. After its new races overcame a number of resistance genes, the genes *Sr2*, *Sr23*, *Sr25*, *Sr33*, *Sr35*, *Sr45*, *Sr47*, and *Sr50* are presently the most efficient for protection against newly evolved races [117]. The QTL-controlling stripe rust resistance in *T. monococcum* was mapped on chromosome 2A (*QYrtm.pau-2A*), whereas the QTL from *T. boeoticum* was mapped on 5A (*QYrtm.pau-5A*). One stripe rust-resistant gene from *T. boeoticum* acc. pau5088 was confirmed to be introgressed in cultivated wheat which was indicated by co-introgression of *T. boeoticum* sequences linked to stripe rust-resistant QTL, *QYrtb.pau-5A* [118].

For stripe (yellow) rust resistance, 95 genes have been characterized and formally named [105, 114, 115]. However, most of these genes have been rendered ineffective with emergence of virulent races around the globe with exception of a few combinations, such as the combination of *Yr5* and *Yr15* that remain effective worldwide. At Punjab Agricultural University, Ludhiana, India, about 200 accessions of *T. monococcum* and *T. boeoticum* were screened for leaf rust and stripe rust resistance for several years and we found that all the *T. monococcum* accessions, most of the *T. boeoticum* and a few *T. urartu* accessions, were completely resistant to leaf rust. However, a lot of variation was observed for stripe rust resistance. Leaf and stripe rust resistance genes have also been introgressed from diploid species *Ae. umbellulata* and *Ae. caudata* using *T. durum* as bridging species [118, 119].

Similarly, 80 *Lr* resistance genes have been genetically characterized and documented [115]. Out of these, *Lr1*, *Lr3*, *Lr10*, and *Lr20* have been commonly deployed in wheat cultivars [120]. Generally, ASR genes are rendered ineffective with continual emergence of new virulent races of rust pathogens through mutation and recombination [121]. It has been well documented through cloning of 11 race-specific genes in wheat (*Sr22*, *Sr33*, *Sr35*, *Sr45*, *Sr50*, *Yr5*, *Yr10*, *Lr1*, *Lr10*, *Lr21*, and *Lr22*) that these genes encode NLR proteins [122–126].

Till date, only seven race non-specific APR genes have been genetically characterized and formally designated in wheat namely *Sr2/Yr30*, *Lr34/Yr18/Sr57/Pm38*, *Lr46/Yr29/Sr58/Pm39*, *Lr67/Yr46/Sr55/Pm46*, *Lr68*, *Sr56*, and *Yr36* [127–133]. Cloning of the APR genes *Yr36*, *Lr34/Yr18/Sr57/Pm38* and *Lr67/Yr46/Sr55/Pm46* has revealed the roles of cytoplasmic protein kinase, adenosine triphosphate (ATP)-binding cassette transporter, and hexose transporter, respectively in mediating resistance [134–136].

Growing resistant cultivars is the most cost-effective strategy for tackling PM. To date, 70 PM resistance genes have been formally cataloged; most of these provide race-specific resistance in wheat [114, 115]. It is desirable to know the virulence pattern of isolates to generate effective combinations of race-specific resistance genes [137]. More effective method would be deployment of combinations of race non-specific resistance genes is a promising method. As discussed above in the section for rust resistance, only three race non-specific resistance genes have been identified, out of which two pleiotropic genes (*Lr34/Yr18/Sr57/Pm38* and *Lr67/Yr46/Sr55/Pm46*) have been cloned [135, 136].

Genetic resistance to FHB is mainly quantitative and is controlled by multiple moderate to minor genes [138]. Although genetic resistance is the most cost-effective method, it is hard to accomplish in commercial cultivars due to its complex behavior. This complexity is further enhanced by various resistance mechanisms, e.g., invasion (type I), fungal spread (type II), toxin accumulation (type III), kernel infection (type IV) and yield reduction (type V) [139]. FHB resistance also displays significant correlations with heading, plant height, and anther extrusion of the wheat plant [140]. To date, seven genetic loci designated as *Fhb1*, *Fhb2*, *Fhb4* and *Fhb5* from wheat, and *Fhb3*, *Fhb6* and *Fhb7* from wild relatives, have been formally named as FHB resistance genes [141]. The cultivars Sumai 3 from China and Frontana from Brazil have been identified as sources of moderate resistance to FHB.

Karnal bunt is among the few quarantine diseases that restrict free trade among countries due to quarantine regulations [142]). Resistance to Karnal bunt has been reported in durum wheat (*Triticum turgidum*), common wheat, *Aegilops*, rye and barley under artificial conditions [143, 144]. Susceptibility of *T. aestivum* to Karnal bunt might be due to presence of an additional D genome [145, 146]. Sharma et al. [147] at PAU developed high yielding Karnal bunt resistant wheat lines by introgression of Karnal bunt resistance from KBRL 22 into the background of high yielding PBW343. Studies on deciphering genetics of resistance have indicated the presence of quantitative rather than qualitative resistance [145, 146, 148]. Fuentes-Davila et al. [145] suggested six genes, designated Kb1, Kb2, Kb3, Kb4, Kb5, and Kb6, while Villareal et al. [149] postulated a minimum of three genes for resistance. Studies on deciphering genetics of resistance have indicated the presence of quantitative rather than qualitative resistance [145, 148].

For loose smut, the majority of genetic studies carried out thus far have demonstrated simple inheritance with one, two or three major genes in hexaploid wheat controlling resistance to several races of *U. tritici*. The first four loose smut resistance genes *Ut1* to *Ut4* were named based on segregation of avirulence in *U. tritici* [150, 151]. Genes *Ut1* and *Ut3* have no chromosome assignment. Based on pedigree, the gene symbol *Ut2* was assigned to the resistance gene on chromosome 6A to race T19 [152]. *Ut4* associated with the Thatcher derived differential line TD12A, was located on chromosome 7B [153, 154]. *Ut5* was located on chromosome 2BL [155], *Ut6* was initially reported on chromosome 5B by Kassa et al. [156] which was later validated by Knox et al. [153]. A gene located to chromosome 7A by Dhitaphichit et al. [157] was subsequently named *Ut7* [153]. Knox et al. further identified genes *Ut8* on chromosome 3A, *Ut9* on chromosome 6B and *Ut10* on chromosome 6D. Several studies revealed the additive nature of resistance genes, while in some cases, duplicate complementary action of multiple genes was also implicated [158].

Finally, the genetic resistance to wheat blast at the seedling stage follows a gene-for-gene interaction model [159] and five resistance genes namely *Rmg2, Rmg3, Rmg7, Rmg8*, and *RmgGR119* have been identified in wheat against the *Magnaporthe oryzae* pathotype *triticum* [160–164].

Various molecular markers have been widely used to tag and map resistance genes in wheat; however, SSRs have emerged as the choice of marker in gene-mapping studies. These markers can be strategically used for selection of desirable gene combinations along with phenotypic assays. Wheat has more than 3000 SSR markers mapped so far [165]. Molecular markers can be used for alien gene transfers and understanding the mechanism of gene transfer. Such markers ensure selection of a target gene based on the presence of the linked genotype. The success of selection depends on the close genetic association and robustness of a given marker across different genetic backgrounds. At PAU, a number of genes/QTLs have been mapped for different wheat diseases including stripe rust, cereal cyst nematode, and Karnal bunt. Two QTLs, one each in *T. monococcum* acc. pau14087, and *T. boeoticum* acc. pau5088, were detected for resistance in the RIL population. The QTL in *T. monococcum* mapped on 2A in a 3.6 cM interval between *Xwmc407* and *Xwmc170*, whereas the QTL from *T. boeoticum* mapped on 5A in 8.3 cM interval between *Xbarc151* and *Xcfd12* [166–168].

9.2 Insect-pests

In the last 50 years or so, the HPR concept has been extended to insect-host interactions. As a result, insect resistant cultivars are now in the picture. The variables, both biotic and abiotic which play a major role in deciding the plant reaction to pest, along

Insect-pest	Order	Gene(s)	Category	References
Aceriatosichella	Acari	*Cmc*(4)	Ab	[169]
Cephuscinctus	Hymenoptera	*Qssmsub* (2); QTL	Ab, Ax, Tol	[170, 171]
Diuraphisnoxia	Hemiptera	*Dn* (10); QTL	Ab, Ax, Tol	[172, 173]
Mayetiola destructor	Diptera	*H* (>33)	Ab	[174]
Schizaphis graminum	Hemiptera	*Gb* (>10); QTL	Ab, Ax, Tol	[175]
Sitodiplosis mosellana	Diptera	*Sm* (1); QTL	Ab	[176, 177]

Ab: antibiosis; Ax: antixenosis; QTL: quantitative trait loci; Tol: tolerance.

Table 3.
Genes identified for insect resistance in wheat and their respective categories.

with mechanisms and categories of resistance are now better understood. Drawing analogy from plant-pathogen interactions, pest-host relationships are now being viewed as (susceptible plant) and incompatible (resistant plant) interactions [74].

Deployment of insect resistance genes in wheat along with other field crops has increased steadily over the years from mid 60s. Marker assisted selection (MAS) and breeding has sped up the process of identification of resistance loci and QTLs and understanding of the mechanisms governing the resistance. **Table 3** depicts the genes identified for insect resistance in wheat and their respective categories.

10. Key challenges

Wheat is an allopolyploid resulted from interspecific or intergeneric hybridization of two or more genomes from different species. Being one of the most consumed and cultivated crop globally, it is under continuous exposure to a large variety of parasite species and strains, many of which have the ability to move around the globe. Long-term co-evolution between plants and their pathogens has equipped plants with a sophisticated multi-layered immune system to guard themselves against pest and pathogens [178]. Despite this, there are a few important challenges which are required to be addressed for effectively mitigating with different biotic stresses in wheat:

1. New strains of pathogens like the rusts continue to evolve rapidly. It is well documented that the rust pathogens have great pathogenic variability and the frequent emergence of new virulent strains that overcome resistance genes present in cultivated wheat varieties has hindered efforts to achieve durable resistance to these pathogens.

2. The complex nature of plant–parasite interactions can be overwhelming while breeding for disease resistance in wheat. The standard models of plant pathology i.e. gene for gene model and the expanded model of plant immunity do not elucidate plant immunity and parasite adaptation explicitly in such natural interactions.

3. The bread wheat (*Triticum aestivum*) genome is one of the most challenging plant genomes to study. It is highly repetitive (~85%) and approximately 15.4–15.8 Gbp in size [179]. Much of the desirable genetic diversity is present in the wild relatives of wheat, both in progenitors and non-progenitor species. The genomic complexity of bread wheat and various hybridization barriers hinder the potential use of resistance alleles present in that germplasm.

4. Despite the versatility of transgenic technology with unlimited scope for application in wheat resistance breeding, it has faced increasing public dissent especially against its use in food crops. Other issues include rigorous risk assessments of crop, which are time-consuming and cost-intensive. Such modifications lead to integration of transgenes randomly into plant genomes along with their selection marker genes. Due to which, there is a possibility of pleiotropic effects, potential silencing and varied gene expression in modified plants

5. Traditional map based/positional cloning is not viable for target genes derived from wild relatives of wheat and which are located in introgressed genome segments that do not recombine with wheat chromatin. Applying this strategy on genes that are located in centromeric regions is also extremely challenging (low recombination rates there).

6. The foremost challenge in breeding against insect pests is finding sources with reasonable levels of resistance against the pest. Secondly another major hurdle is the difference between resistance at field and protected conditions, since evaluation is carried out in protected conditions, results vary when evaluation is carried out *in vivo*. Lack of efficient evaluation and selection tools against insects also hinders the insect resistance breeding. Finally, transfer of resistance is often accompanied by linkage drag which sometimes becomes cumbersome to break.

11. Conclusion and future prospects

Genetic control is considered as the most effective and environmentally friendly strategy to control rust disease and involves breeding effective disease resistance genes into wheat cultivars. Many rust resistance genes have been identified genetically, and introgression into wheat lines is increasingly being facilitated by the development of robust molecular markers. However, the massive and complex genome of wheat poses major challenges for the isolation of individual genes. As revealed by the increasing number of newly available whole genome sequences and the more precise bioinformatic pipelines developed for identifying NLR genes, the number of NLR genes varies greatly between species. Based on an analysis of the IWGSC RefSeq v1.0 assembly, a total of 3,400 full-length NLR loci have been documented [180]. The approaches for identifying effective resistance genes therefore, must consider both classical R-genes (immune receptor class genes) as well as other novel classes that may operate via different mechanisms.

Cloning of the genes that controlling resistance to rust pathogens will significantly advance our understanding of the molecular basis underlying expression of disease resistance in wheat. Only a small number of rust resistance genes have been cloned and had their molecular functions studied (**Table 2**). To overcome the limitations of the map-based cloning strategy in the large genome of wheat, alternative approaches were developed and validated by the rapid cloning of several genes using Target-sequence Enrichment and Sequencing (TEnSeq) pipelines. These include MutRenSeq (Mutagenesis and the Resistance gene Enrichment and Sequencing), AgRenSeq (Association genetics with R gene enrichment Sequencing), MutChromSeq (Mutgenesis Chromosome flow sorting and short-read Sequencing), and TACCA (Targeted Chromosome based Cloning via long-range Assembly). The common component among all these approaches is the intent to reduce the genome complexity prior to the use of next generation sequencing (NGS). Such insight into the molecular mechanisms will be the foremost step towards the functional characterization of the wheat-rust interaction and allow engineering of new resistance by exploiting novel techniques like allele mining and genome editing. Also, approaches like TILLING (Targeting Induced Local Lesions IN Genomes) can be adopted for more precise and efficient characterization of the function of targeted wheat genes for different fungal and bacterial diseases.

The rich genetic diversity available in wheat is a source of numerous novel alleles for both disease resistance and tolerance to abiotic stress. However, there is still a huge gap in characterization of the available genetic resources and their utilization in breeding programs. Over the years, traditional breeding strategies have successfully incorporated novel alleles into elite germplasm, which has significant impacts on production globally. Use of advanced technologies, marker-assisted selection (MAS), genomic selection, transgenics and genome editing will help to increase the efficiency of wheat breeding for biotic stress resilience around the world.

To escape the boom and bust cycle, resistance gene stewardship and deployment strategies such as gene pyramiding, gene stacking (transfer of gene cassettes) could prove to be effective against deadly diseases of wheat (rusts, blight). It is widely reported and agreed upon fact that the most effective and durable means for genetic control of wheat rusts is the use of combinations of multiple broadly effective ASR and APR genes. Using this, the desirable combinations of effective resistance genes can be combined and transformed into wheat as gene cassettes or stacks. This can result in faster improvements in disease resistance of current high-yielding varieties. Also, the advancements in R-gene cloning pipeline like TEnSeq will provide many more tools for MAS in wheat breeding as well as the raw gene sequences to pursue gene stacking (via transgenic gene cassettes). Combining with advances in identifying genetic variation in rust *Avr* genes, these new tools will lead to more effective deployment strategies to maximize resistance durability.

Genomic selection (GS) is considered one of the best strategies for selection of multiple minor-effect loci in comparison with MAS. Using GS, a training population (after phenotyping and genotyping) is used to standardize a prediction model, which is further used to predict breeding values, thus enabling selection of candidates prior to phenotyping [181]. Recent studies have reported that greater genetic gains can be obtained by using genomic selection than by using MAS [182] and phenotypic selection [183].

More recently, genome editing has emerged as a prominent new plant breeding technique, which involves targeted modification of a native DNA sequence. For instance, it has been observed that a single amino acid substitution (Arg144Gly) in a hexose transporter in wheat results in the gene *Lr67* conferring resistance. This substitution evolved recently after common wheat polyploidization. Introduction of the *Lr67* transgene into barley conferred seedling and adult plant resistance to the barley leaf rust pathogen [184, 185]. The orthologue sequence of *Lr67* exists in the barley genome; hence altering the Arg144Gly by genome editing would be expected to produce resistance to rust in barley. Similarly, a number of homologs/orthologues of the isolated genes exist in related species. Isolating a rust resistance gene from other related species thus can provide deeper insight into rust resistance in the wheat.

Therefore, under a changing global climate, it is of paramount importance to breed for durable and broad-spectrum disease resistance in wheat at a faster pace to reduce losses from attack by rapidly evolving new virulent pathogenic races. Moreover, this would lead to reduction of the use of agrochemicals (fungicides), escaping environmental and human health hazards, an essential component of modern sustainable crop production systems.

Author details

Harmeet Singh Bakala*, Kamalpreet Singh Mandahal, Ankita, Loveleen Kaur Sarao and Puja Srivastava
Punjab Agricultural University, Ludhiana, India

*Address all correspondence to: harmeet-pbg@pau.edu

References

[1] Gupta P K, Mir R R, Mohan A and Kumar J. Wheat Genomics: Present Status and Future Prospects. International Journal of Plant Genomics Volume 2008. p. 1-36

[2] Bertholdsson NO. Early vigor and Allelopathy- Two useful traits for enhancing barley and wheat competitiveness against weeds. Weed Research. 2005;**45**:94-102

[3] Figueroa M, Hammond-Kosack KE, Solomon PS. A review of plant diseases—a field perspective. Molecular Plant Pathology. 2017;**19**:1523-1536

[4] Weise MV, editor. Compendium of wheat diseases. 2nd ed. American Phytopathology Society,St. Paul 1987

[5] Wellings CR. Global status of stripe rust: a review of historical and current threats. Euphytica 2011;**179**:129-141

[6] Plank J. E. van der. Plant diseases: epidemics and control. 1963. New York: Academic. 349 pp

[7] Priyamvada, Saharan M S, Tiwari R. Durable resistance in wheat. International Journal of Genetics and Molecular Biology. 2011; **3(8)**:108-114

[8] Hovmøller SM, Sørensen CK, Walter S, Justesen A F. Diversity of Puccinia striiformis on Cereals and Grasses. Annual Review Phytopathology.2011; **49**:197-217

[9] Singh R P, Hodson DP, Huerta-Espino J et al. The 61 Emergence of Ug99 Races of the Stem Rust Fungus is a Threat to World Wheat Production. Annual Review Phytopathology. 2011; **49**:465-482

[10] Singh RP, William HM, Huerta-Espino J, Rosewarne G. Wheat rust in Asia: meeting the challenges with old and new technologies. New dimensions for a diverse planet. In: Proceedings of the 4th International Crop Science Congress, 26 Sep–1 Oct 2004, Brisbane

[11] Rizwan S, Ahmad I, Ashraf M, Mirza JI, Sahi GM, Rattu AR, Mujeeb-Kazi A. Evaluationof synthetic hexaploid wheats and their durum parents for stripe rust resistance. Rev MexFitopatol 25:152-160

[12] Roelfs AP, Singh RP, Saari EE. Rust diseases of wheat: concepts and methods of diseasemanagement. CIMMYT, Mexico DF. http://hdl.handle.net/10883/1153

[13] Singh RP, Hodson DP, Huerta-Espino J, Jin Y, Njau P. Will stem rust destroy theworld's wheat crop? Advances in Agronomy. 2008;**98**: 271-309

[14] Cruz CD, Valent B. Wheat blast disease: danger on the move. Tropical Plant Pathology. 2017;**42(3)**:210-222

[15] Islam MT, Croll D, Gladieux P. Emergence of wheat blast in Bangladesh was caused by a South American lineage of *Magnaporthe oryzae*. BMC Biology. 2016; **14**:84

[16] Tembo B, Mulenga RM, Sichilima S, Msiska KK, Mwale M, Chikoti PC. Detection and characterization of fungus (*Magnaportheoryzae pathotype Triticum*) causing wheat blast disease on rain-fed grown wheat (*Triticum aestivum* L.) in Zambia. Plosone. 2020;**15**(9): e0238724

[17] Mitra M. A new bunt on wheat in India. Annals of Applied Biology. 1931; **18**:178-179

[18] Rush CM, Stein JM, Bowden RL, Riemenschneider R, Boratynski T, Royer MH. Status of Karnal bunt of wheat in United States 1996-2004. Plant Disease. 2005;**89**:212-222

[19] Dhaliwal HS, Singh H, Singh KS, Randhawa HS. Evaluation and cataloguing of wheat germplasm for disease resistance and quality. In: Damania AB ed. Biodiversity and wheat improvement. Wiley, London,1993. pp 123-140

[20] Dhaliwal HS, Singh H. Breeding for resistance to bunts and smuts: Indian scenario. In:Proceedings bunts and smuts of wheat: an international symposium. North Carolina, North American Plant Protection Organization, Ottawa 1997, pp 327-347

[21] Katare S, Singh B,Patil SD, Tiwari R, Jasrotia P, Saharan MS, Sharma I. Evaluation of new insecticides for management of foliar aphid complex in wheat. Indian Journal of Entomology 2015;79(2):185-190.

[22] Kieckhefer RW, Gellner JL. Yield losses in winter wheat caused by low-density cereal aphid populations. Agron J 1992; 84:180-183.

[23] Kieckhefer R W, Gellner, JL, Riedell WE. Evaluation of the aphid-day standard as a predictor of yield loss caused by cereal aphids. Agronomy J 1995;87(5):785-788.

[24] RyanJD, Dorschner KW, Eikenbary RD, Johnson RC. Drought/greenbug interactions: photosynthesis of greenbug resistant and susceptible wheat. Crop Sci.1987;27:283-288.

[25] BurtonRL. Effect of greenbug (Homoptera: Aphididae) damage on root and shoot biomass of wheat seedlings. J. Econ. Entomol.1986; 79633636

[26] Capinera JL, editor. Encyclopedia of entomology, 2nd ed. Springer, Netherlands; 2008. 4242 p.

[27] Beres BL, Dosdall LM, Weaver DK. Biology and integrated management of wheat stem sawfly and the need for continuing research. Can Entomol. 2011;143:105-125. DOI: 10.4039/n10-056

[28] Andersen E J, Ali S, Byamukama E, Yen Y, Nepal P. Disease resistance mechanisms in plants. Genes. 2018;**9**:339.

[29] Flor H H. Current status of the gene-for-gene concept. Annual Review of Phytopathology. 1971; **9**:275-296

[30] Jones JDG, Vance RE, Dangl JL. 2016. Intracellular innate immune surveillance devices in plants and animals. Science. 2016;**354:**aaf6395-1-8

[31] Dodds PN, Rathjen JP. Plant immunity: towards an integrated view ofplant–pathogen interactions. Nature Reviews Genetics. 2010;**11**:539-548

[32] Singla J,Krattinger S G.Biotic Stress Resistance Genes in Wheat. Encyclopedia of food grains. 2016; 388-392

[33] Huang L, Brooks S A, Li W, Fellers J P, Trick H N, Gill B S. Map-based cloning of leaf rust resistance gene Lr21 from the large and polyploid genome of bread wheat. Genetics. 2003;**164**:655-664.

[34] FeuilletC, Travella S, Stein N, Albar L, Nublat A, Keller B. Map-based isolation of the leaf rust disease resistance gene Lr10 from thehexaploid wheat (*Triticum aestivum* L.) genome. Proc. Natl. Acad. Sci. U.S.A. 2003;**100**: 15253-15258.

[35] Cloutier S, McCallum B D, Loutre C, Banks T W, Wicker T, Feuillet C (2007). Leaf rust resistance gene Lr1, isolated from bread wheat (Triticumaestivum L.) is a member of the large psr567 gene family. Plant Molecular Biology. 2007;**65**:93-106.

[36] Krattinger S G, Lagudah E S, Spielmeyer W, Singh R P,

Huerta-EspinoJ, McFadden H. A putative ABC transporter confers durableresistance to multiple fungal pathogens in wheat. Science. 2009;**323**:1360-1363.

[37] Moore J W, Herrera-Foessel S, Lan C, Schnippenkoetter W, AyliffeM,Huerta-Espino J. A recently evolved hexose transporter variantconfers resistance to multiple pathogens in wheat. Nature Genetics. 2015;**47**:1494-1498.

[38] Thind A K, Wicker T, Simkova H, Fossati D, Moullet O, Brabant C. Rapid cloning of genes in hexaploid wheat using cultivar-specific longrangechromosome assembly. Nature Biotechnology. 2017;**35**: 793-796.

[39] Fu D, Uauy C, Distelfeld A, Blechl A, EpsteinL, Chen X. Akinase-START gene confers temperature-dependent resistance to wheat striperust. Science. 2009;**323**: 1357-1360.

[40] Marchal C, Zhang J, Zhang P, Fenwick P, Steuernagel B, Adamski N M. BED-domain-containing immune receptors confer diverse resistancespectra to yellow rust. Nature Plants. 2018;**4**:662-668.

[41] Klymiuk V, Yaniv E, Huang L, Raats D, Fatiukha A, Chen S S.Cloning of the wheat Yr15 resistance gene sheds light on the plant tandemkinase-pseudokinase family. Nature Communication. 2018;**9**:3735

[42] Zhang C, Huang L, Zhang H, Hao Q, Lyu B, Wang M. Anancestral NB-LRR with duplicated 3'UTRs confers stripe rust resistance inwheat and barley. Nature Communication. 2019;**10**:4023

[43] Periyannan S, Moore J, Ayliffe M, Bansal U, Wang X, Huang L.The gene Sr33, an ortholog of barley Mla genes, encodes resistance to wheatstem rust race Ug99. Science. 2013;**341**:786-788

[44] Saintenac C, Zhang W, Salcedo A, Rouse M N, Trick H N, Akhunov E. Identification of wheat gene Sr35 that confers resistance to Ug99 stemrust race group. Science. 2013;**341**:783-786.

[45] MagoR, Zhang P, Vautrin S, Simkova H, Bansal U, Luo M C. The wheat Sr50 gene reveals rich diversity at a cereal disease resistance locus.Nature Plants. 2015;**1**:15186.

[46] Steuernagel B, Periyannan SK, Hernandez-Pinzon I, Witek K, Rouse M N,Yu G. Rapid cloning of disease-resistance genes in plants usingmutagenesis and sequence capture. Nature Biotechnology. 2016;**34**:652-655.

[47] Zhang W, Chen S, Abate Z, Nirmala J, Rouse M N,DubcovskyJ. Identification and characterization of Sr13, a tetraploid wheat gene that confersresistance to the Ug99 stem rust race group. Proc. Natl. Acad. Sci. U.S.A. 2017;**114**:E9483–E9492.

[48] Chen S, Zhang W, Bolus S, Rouse M N, DubcovskyJ.Identification and characterization of wheat stem rust resistance gene Sr21effective against the Ug99 race group at high temperature. PlosoneGenetics. 2018;**14**:e1007287.

[49] Arora S, Steuernagel B, Gaurav K, Chandramohan S, Long Y M, Matny O. Resistance gene cloning from a wild crop relative by sequencecapture and association genetics. Nature Biotechnology. 2019;**37**:139-143

[50] Chen S, Rouse M N, Zhang W, Zhang X, Guo Y, Briggs J.Wheat gene Sr60 encodes a protein with two putative kinase domains thatconfers resistance to stem rust. New Phytology. 2019;**225**:948-959.

[51] Yahiaoui N, Srichumpa P, Dudler R, Keller B. Genome analysis atdifferent

ploidy levels allows cloning of the powdery mildew resistance genePm3b from hexaploid wheat. Plant Journal. 2004;**37**:528-538

[52] Hurni S. Rye Pm8 and wheat Pm3 are orthologous genes and showevolutionary conservation of resistance function against powdery mildew.Plant Journal. 2013;**76**:957-969

[53] Sanchez-Martin J. Rapid gene isolation in barley and wheat by mutantchromosome sequencing. Genome Biology. 2016;**17**: 221

[54] Xing L P. Pm21 from Haynaldiavillosa encodes a CC-NBS-LRR proteinconferring powdery mildew resistance in wheat. Molecular Plant.2018; **11**:874-878

[55] He H G. Pm21, encoding a typical CC-NBS-LRR protein, confersbroad-spectrum resistance to wheat powdery mildew disease. Molecular Plant. 2018;11,:879-882

[56] Zou S H, Wang H, Li Y W, Kong Z S, Tang D Z. The NB-LRR genePm60 confers powdery mildew resistance in wheat. New Phytology. 2018;**218**:298-309

[57] Rawat N, Pumphrey M, Liu S.(2016) Wheat *Fhb1* encodes a chimeric lectin with agglutinin domains and a pore-forming toxin-like domain conferring resistance to Fusarium head blight. Nature Genetics. 2016;**48:**1576-1580

[58] Paudel B, Zhuang Y, Galla A, Dahal S, Qiu Y, Ma A, Raihan TYenY.*WFhb1-1* plays an important role in resistance against Fusarium head blight in wheat. Scientific Reports. 2020; 10:7794

[59] Joshi A. Map-based cloning of the Hessian fly resistance gene H13 in Wheat [thesis]. Kansas State University; 2018.

[60] Dodds PN, Lawrence GJ, Ellis JG. Six amino acid changes confined to the leucine-rich repeat beta-strand/beta-turn motif determine the difference between the P and P2 rust resistance specificities in flax.The Plant Cell. 2001;**13**:163-178.

[61] Yahiaoui N, Brunner S, Keller B. 2006. Rapid generation of newpowdery mildew resistance genes after wheat domestication. The Plant Journal. 2006;**47**:85-98.

[62] Ashfield T, Redditt T, Russell A, Kessens R, Rodibaugh N, GallowayL, Kang Q, Podecheti R, Innes RW. Evolutionary relationshipof disease resistance genes in soybean and *Arabidopsis* specific for the *Pseudomonas syringae* effectors *Avr B*and *Avr Rpm1*. Plant Physiology. 2014;**166**:235-251

[63] Tommasini L, Yahiaoui N, Srichumpa P, Keller B. 2006. Development of functional markers specific for seven *Pm3* resistance alleles and their validation in the bread wheat gene pool. Theoretical and Applied Genetics. 2006; **114**:165-175

[64] Lagudah ES, Krattinger SG, Herrera-Foessel S, Singh RP, Huerta-Espino J, Spielmeyer W, Brown-Guedira G, Selter LL, Keller B. Gene-specific markers for the wheat gene *Lr34/Yr18*/Pm38 which confers resistance to multiple fungal pathogens. Theoretical Applied Genetics. 2009;**119**:889-898

[65] Panda N, Khush GS. Host Plant Resistance to Insects.Wallingford, UK: CABI/IRRI. 1995. 431 p.

[66] Smith CM. Plant Resistance to Arthropods: Molecular and Conventional Approaches. Dordrecht, The Netherlands: Springer. 2005. 423 p.

[67] Sadasivam S, Thayumanavan B. Molecular Host Plant Resistance to Pests. New York:Marcel Dekker. 2003. 479 p.

[68] Smith CM, Liu XM, Wang LJ, Liu X, Chen MS. Aphid feeding activates expression of a transcriptome of oxylipin-based defense signals in wheat involved in resistance to herbivory. J. Chem. Ecol. 2010;36:260-276.

[69] Chen MS. Inducible direct plant defense against insect herbivores: a review. Insect Sci.2008;15:101-114.

[70] Peshin R, Dhawan AK, editors. Integrated Pest Management: Innovation-Development Process New York/Heidelberg: Springer Science + Business Media; 2009; 690 p.

[71] Kessler A, Baldwin IT. Plant responses to insect herbivory: the emerging molecular analysis. Annu. Rev. Plant Biol.2002;53:299-328.

[72] Mithofer A, Boland W. Recognition of herbivory-associated molecular patterns. Plant Physiol.2008;146: 825-831.

[73] Schmelz EA, Engelberth J, Alborn HT, Tumlinson JH, Teal PEA. Phytohormone-based activity mapping of insect herbivore-produced elicitors. Proc. Natl. Acad. Sci. USA2009;106: 653-57.

[74] Kaloshian I. Gene-for-gene disease resistance. bridging insect pest and pathogen defense. J. Chem. Ecol.2004;30:2419-2438.

[75] Sullivan TJ, Rodstrom J, Vandop J, Librizzi J, Graham C. Symbiont-mediated changes in *Lolium arundinaceum*inducible defenses: evidence from changes in gene expression and leaf composition. New Phytol. 2007;176:673-679.

[76] Underwood N, Rausher M. Comparing the consequences of induced and constitutive plant resistance for herbivore population dynamics. Am. Nat. 2002;160:20-30.

[77] Huang J, McAuslane HJ, Nuessly GS. Resistance in lettuce to *Diabroticabalteat a*(Coleoptera: Chrysomelidae): the roles of latex and inducible defense. Environ. Entomol.2003;32:9-16.

[78] Underwood NC, Rausher M, Cook W. Bioassay versus chemical assay: measuring the impact of induced and constitutive resistance on herbivores in the field. Oecologia2002;131:211-219.

[79] Couldridge C, NewburyHJ, Ford-Lloyd B, Bale J, Pritchard J. Exploring plant responses to aphid feeding using a full *Arabidopsis* microarray reveals a small number of genes with significantly altered expression. Bull. Entomol. Res.2007;97:523-532.

[80] Kielkiewicz M. Influence of carmine spider mite TetranychuscinnabarinusBoisd. (Acarida: Tetranychidae) feeding on ethylene production and the activity of oxidative enzymes in damaged tomato plants. In Acarid Phylogeny and Evolution: Adaptation in Mites and Ticks—Proc. IV Symp. Eur. Assoc. Acarol., ed. F Bernini, R Nannelli, G Nuzzaci, E de Lillo;2002; Dordrecht, The Netherlands: Kluwer;2002. p. 389-92

[81] Li Y, Zou J, Li M, Bilgin DD, Vodkin LO. Soybean defense responses to the soybean aphid. New Phytol. 2008; 179:185-195.

[82] Liu X, Bai J, Huang L,Zhu L, Liu X. Gene expression of different wheat genotypes during attack by virulent and avirulent Hessian fly (*Mayetiola destructor*) larvae. J. Chem. Ecol. 2007;33:2171-2194.

[83] Kaloshian I, Walling L. Hemipterans as pathogens. Annu. Rev. Phytopathol.2005;43:491-521.

[84] Smith CM, Boyko EV. The molecular bases of plant resistance and defense responses to aphid feeding: current status. Entomol. Exp. Appl. 2007;122:1-16.

[85] van Eck L, Schultz T, Leach JE, Scofield SR, Peairs FB. Virus-induced gene silencing of *WRKY53* and an inducible *phenylalanine ammonia-lyase* in wheat reduces aphid resistance. Plant Biotechnol. J.2010;8:1023-1032.

[86] Howe GA, Jander G. Plant immunity to insect herbivores. Annu. Rev. PlantBiol.2008;59:41-66.

[87] Zhu L, Liu X, Liu XM, Jeannotte R, Reese J. Hessian fly (*Mayetiola destructor*) attack causes a dramatic shift in carbon and nitrogen metabolism in wheat. Mol. Plant-Microbe Interact.2008;21:70-78.

[88] Lapitan NLV, Li YC, Peng JH, Botha AM. Fractionated extracts of Russian wheat aphid eliciting defense responses in wheat. J. Econ. Entomol. 2007;100:990-999.

[89] Thackray DJ, Wratten SD, Edwards PJ, Niemeyer HM. Hydroxamic acids - potential resistance factors in wheat against the cereal aphids *Sitobionavenae* and *Rhopalosiphumpadi*. Proceedings of 1990 Brighton Pest Control Conference-Pests and Diseases-1990.1991;p215-220.

[90] Gianoli E, Ríos JM, Niemeyer HM. Allocation of a hydroxamic acid and biomass during vegetative development in rye. Acta Agriculture Scandinavica, Section B. Soil and Plant Science. 2000;50:35-39.

[91] Givovich A, Niemeyer HM. Comparison of the effect of hydroxamic acids from wheat on five species of cereal aphids. EntomologiaExperimentalis et Applicata. 1995;74:115-119.

[92] Givovich A, Niemeyer HM. Effect of hydroxamic acids on feeding behavior and performance of cereal aphids on wheat. European Journal of Entomology 1994;91:371-374.

[93] Figueroa C, Simon J, Gallic J, Prunier–letermeN, Briones lM, DedryverC,Niemeyer hM.Effect of host defense chemicals on clonal distribution and performance of different genotypes of the cereal aphid *Sitobionavenae*. Journal of Chemical Ecology. 2004;30(12):2515-2525.

[94] Kaur H, Salh P.SinghB. Role of defense enzymes and phenolics in resistance of wheat crop (Triticum aestivum L.) towards aphid complex. J.PlantInteractions. 2017;12(1):304-311.

[95] Nemacheck JA, Schemerhorn BJ, Scofield SR, Subramanyam S. Phenotypic and molecular characterization of Hessian fly resistance in diploid wheat, *Aegilops tauschii*. BMC Plant Biol. 2019;19(1):439. DOI:10.1186/s12870-019-2058-6

[96] Puthoff DP, Sardesai N, Subramanyam S, Nemacheck JA, Williams CE. Hfr-2, a wheat cytolytic toxin-like gene, is up-regulated by virulent hessian fly larval feeding. Mol Plant Pathol. 2005;6:411-423.

[97] Subramanyam S, Sardesai N, Puthoff DP, Meyer JM, Nemacheck JA, Gonzalo M, Williams CE. Expression of two wheat defense-response genes, Hfr-1 and Wci-1, under biotic and abiotic stresses. Plant Sci. 2006;170:90-103.

[98] FeuilletCS, TravellaNS, teinAlbarL,Nublatand A, Keller B. Map-based isolation of the leaf rust disease resistance gene Lr10 from the hexaploid wheat (*Triticum aestivum* L.) genome. Proc. Natl. Acad. Sci. U.S.A. 2003;100(25):15253-15258.

[99] Botha AM, Lacock L, van Niekerk C, Matsioloko MT, du Preez FB, Loots S, Venter E, Kunert KJ, Cullis CA. Is photosynthetic transcriptional regulation in Triticum aestivum L. cv. 'TugelaDN' a contributing factor for tolerance to Diuraphisnoxia (Homoptera: Aphididae)? Plant Cell

Rep. 2006;25(1):41-54. DOI: 10.1007/s00299-005-0001-9.

[100] Keenn T. Gene-for-gene complementarity in plantpathogen interactions. Annu. Rev. Genet. 1900;24:425-429.

[101] Boyko EV, Smith CM, Thara VK, Bruno JM, Deng Y, Starkey SR, Klaahsen DL. Molecular basis of plant gene expression during aphid invasion: wheat Pto- and Pti-like sequences are involved in interactions between wheat and Russian wheat aphid (Homoptera: Aphididae). J Econ Entomol. 2006;99(4):1430-1445. DOI: 10.1603/0022-0493-99.4.1430.

[102] Appels R, Eversole K, Feuillet C, Keller B, Rogers J, Stein N.Shifting the limits in wheat research and breeding using a fully annotatedreference genome. Science. 2018;**361**:eaar7191.

[103] Jiang J, Friebe B, Gill BS (1994) Recent advances in alien gene transfer in wheat. Euphytica 73:199-212

[104] Friebe B, Jiang J, Raupp WJ, McIntosh RA, Gill BS (1996) Characterization of wheat-alien translocations conferring resistance to diseases and pests: current status. Euphytica 91:59-87

[105] McIntosh RA, Wellings CR, Park RF (1995) Wheat rusts. An atlas of resistance genes. CSIRO Publishing, Melbourne

[106] McIntosh RA, Yamazaki Y, Dubcovsky J, Rogers J, Morris C, Appels R, Xia XC (2013) Catalogue of gene symbols for wheat. In: 12th international wheat genetics symposium, Yokohama, Japan

[107] Lopes, M.S., El-Basyoni, I., Baenziger, P.S., Singh, S., Royo, C., Ozbek, K., et al., 2015b. Exploiting genetic diversity from landraces in wheatbreeding for adaptation to climate change. Journal of Experimental Botany. 66, pp: 3477-3486.

[108] Yang, J. and Liang, Q., 1995. Yinchun 3 wheat germplasm with high protein content and resistance to drought. Crop Genetic Resources, 1, pp: 44.

[109] Li, X., Sun, F., Guo, B., Liu, L. and Pang, C., 1997. Evaluation of abiotic stress resistance in hebei winter wheat (*Triticum aestivum*) genetic resources.*Wheat Information Service*, 85, pp: 1-6.

[110] Reynolds, M.P. and Borlaug, N.E., 2006. Impacts of breeding on international collaborative wheat improvement. *J Agric Sci*., 144, pp: 3

[111] Reitz, L. P.and Salmon, S. C. 1968. Origin, history, and use of Norin 10 Wheat1. Crop Sci., 8, pp: 686-689.

[112] Dhaliwal HS, Singh Harjit, William M (2002) Transfer of rust resistance from Aegilops ovata into bread wheat (Triticum aestivum L.) and molecular characterization of resistant derivatives. Euphytica 126:153-159

[113] Dhaliwal HS, Chhuneja P, Gill RK, Goel RK, Singh H (2003) Introgression of disease resistance genes from related species into cultivated wheats through interspecific hybridization. Crop Improv 29:1-18

[114] McIntosh RA, Dubcovsky J, Rogers J, Morris C, Appels R, Xia XC (2016) Catalogue of gene symbols for wheat: 2015-16 supplement.

[115] McIntosh RA, Dubcovsky J, Rogers JW, Morris C, Xia CX (2017) Catalogue of gene symbols for wheat: 2017 supplement.

[116] Singh RP, Hodson DP, Jin Y, Huerta EJ, Kinyua M, Wanyera R, Njau P, Ward RW (2006) Currentstatus, likely migration and strategies to

mitigate the threat to wheat production from race Ug99(TTKS) of stem rust pathogen. CAB Rev 1:54

[117] Singh RP, Hodson DP, Jin Y, Lagudah ES, Ayliffe MA et al (2015) Emergence and spread of newraces of wheat stem rust fungus: continued threat to food security and prospects of geneticcontrol. Phytopathology 105:872-884

[118] Chhuneja P, Kaur S, Garg T, Ghai M, Kaur S, Prashar M, Bains NS, Goel RK, Keller B, Dhaliwal HS, Singh K (2008a) Mapping of adult plant stripe rust resistance genes in diploid a genome wheat species and their transfer to bread wheat. Theor Appl Genet 116:313-324

[119] Riar AK, Kaur S, Dhaliwal HS, Singh K, Chhuneja P (2012) Introgression of a leaf rust resistance gene from Aegilops caudata to bread wheat. J Genet 91:155-161

[120] Dakouri A, McCallum BD, Radovanovic N, Cloutier S. Molecular and phenotypic characterization of seedling and adult plant leaf rust resistance in a world wheat collection. Molecular Breeding. 2013;**32**:663-677

[121] Randhawa MS, Singh RP, Lan C. Interactions among genes *Sr2*/*Yr30*, *Lr34*/*Yr18*/*Sr57* and *Lr68* confer enhanced adult plant resistance to rust diseases in common wheat (*Triticum aestivum*L.) line 'Arula. Australian Journal of Crop Science. 2018;**12**: 1023-1033

[122] Ellis JG, Lagudah ES, Spielmeyer W, Dodds PN. The past, present and future of breeding rust resistant wheat. Frontier Plant Science. 2014;**5**:641.

[123] Mago R, Zhang P, Vautrin S. The wheat *Sr50* reveals a rich diversity at a cereal disease resistance locus. Nature Plants. 2015;**1**:15186.

[124] Steuernagel B, Periyannan SK, Hernandez-Pinzon I. Rapid cloning of disease-resistance genes in plants using mutagenesis and sequence capture. Nature Biotechnology. 2016;**34**:652-655

[125] Thind AK, Wicker T, Simkova H, Fossati D. Rapid cloning of genes in hexaploidy wheat using cultivar-specific long-range chromosome assembly. Nature Biotechnology. 2017;**35**:793-796.

[126] Marchal C, Zhang J, Zhang P, Fenwick P. BED-domain containing immune receptors confer 2 diverse resistance spectra to yellow rust. Nature Plants. 2018;**4**:662.

[127] Bansal U, Bariana H, Wong D, Randhawa M, Wicker T, Hayden M, Keller B. Molecular mapping of an adult plant stem rust resistance gene *Sr56* in winter wheat cultivar Arina. Theoretical Applied Genetics. 2014;**127**:1441-1448.

[128] Dyck PL. The association of a gene for leaf rust resistance with the chromosome 7D suppressor of stem rust resistance in common wheat. Genome. 1987;**29**:467-469

[129] Hare RA, McIntosh RA. Genetic and cytogenetic studies of durable adult-plant resistances in 'Hope' and related cultivars to wheat rusts. Z Planzenzuchtung. 1979;**83**:350-367

[130] Herrera-Foessel SA, Lagudah ES, Huerta-Espino J. New slow-rusting leaf rust and stripe rust resistance genes *Lr67* and *Yr46* in wheat are pleiotropic or closely linked. Theoretical Applied Genetics. 2011;**122**:239-249

[131] Herrera-Foessel SA, Singh RP, Huerta-Espino J.*Lr68*: a new gene conferring slow rusting resistance to leaf rust in wheat. Theoretical Applied Genetics. 2012; **124**:1475-1486

[132] Singh RP, Mujeeb-Kazi A, Huerta-Espino J.*Lr46*: a gene conferring slow rusting resistance to leaf rust in

wheat. Phytopathology. 1998;**88**: 890-894

[133] Uauy C, Brevis JC, Chen X, Khan I, Jackson L, Chicaiza O, Distelfeld A, Fahima T, Dubcovsky J. High-temperature adult-plant (HTAP) stripe rust resistance gene *Yr36* from *Triticum turgidum* ssp. *Dicoccoides* is closely linked to the grain protein content locus *Gpc-B1*. Theoretical Applied Genetics. 2005;**112**:97-105

[134] Fu D, Uauy C, Distelfeld A, Blechl A, Epstein L, Chen X, Sela H, Fahima T, Dubcovsky J (2009) A kinase-START gene confers temperature dependent resistance to wheat stripe rust. Science. 2009; **323**:1357-1360

[135] Krattinger SG, Lagudah ES, Spielmeyer W, Singh RP, Huerta-Espino J, McFadden H, Bossolini E, Selter LL, Keller B. A putative ABC transporter confers durable resistance to multiple fungal pathogens in wheat. Science. 2009;**323**:1360-1363.

[136] Moore JW, Herrera-Foessel S, Lan C. A recently evolved hexose transporter variant confers resistance to multiple pathogens in wheat. Nature Genetics. 2015;**47**:1494-1498.

[137] Wang ZL, Li LH, He ZH, Duan XY, Zhou YL, Chen XM, Lillemo M, Singh RP, Wang H, Xia XC(2005) Seedling and adult plant resistance to powdery mildew in Chinese bread wheat cultivarsand lines. Plant Dis 89:457-463

[138] Singh RP, Singh PK, Rutkoski J. Disease impact on wheat yield potential and prospects of genetic control. Annual Review Phytopathology. 2016;**54**:303-322

[139] Mesterhazy A, Bartok T, Kaszonyi G, Varga M, Toth B, Varga J. Common resistance to different *Fusarium* spp. causing *Fusarium* head blight in wheat. European Journal of Plant Pathology. 2005;**112**:267-281

[140] Buerstmayr H, Adam G, Lemmens M. Resistance to head blight caused by *Fusarium* spp. Inwheat. In: Sharma I (ed) Disease resistance in wheat. CABI, Wallingford. 2012; pp 236-276

[141] Guo J, Zhang X, Hou Y, Cai J, Shen X, Zhou T, Xu H, Ohm HW, Wang H, Li A, Han F, Wang H, Kong L. High-density mapping of the major FHB resistance gene *Fhb7* derived from *Thinopyrum ponticum* and its pyramiding with *Fhb1* by marker-assisted selection. Theoretical Applied Genetics. 2015; **128**:2301-2316

[142] Rush CM, Stein JM, Bowden RL, Riemenschneider R, Boratynski T, Royer MH. Status of Karnal bunt of wheat in the United States 1996 to 2004. Plant Disease. 2005;**89**:212-223.

[143] Dhaliwal HS,Navarete MR, Valdez JC. Scanning electron microscope studies of penetration mechanism of *Tilletia indica* in wheat spikes. Review Mexican Fitopathology. 1988;**7**:150-155.

[144] Gill KS, Aujla SS, Sharma I. Karnal Bunt and Wheat Production. Punjab Agricultural University Ludhiana, India. 1993; pp. 1-153.

[145] Fuentes-Davila G, Rajaram S, Singh G. Inheritance of resistance to Karnal bunt (*Tilletiaindica*Mitra) in bread wheat (*Triticum aestivum* L). Plant Breeding. 1995; **114**:252-254

[146] Sharma I, Bains NS, Singh K, Nanda GS. Additive genes at nine loci govern Karnal bunt resistance in a set of common wheat cultivars. Euphytica. 2005;**142**:301-307.

[147] Sharma I, Bains NS, Nanda GS.Inheritance of Karnal bunt free trait in bread wheat. Plant Breeding. 2004; **123**:96-97.

[148] Sharma I, Bala R, Kumar S, Bains NS. Development of near isogenic lines (NILS) using backcross method of breeding and simultaneous screening against Karnal bunt disease of wheat. Journal Applied Natural Science. 2016;**8**(3)1138-1145.

[149] Villareal RL, Fuentes- Davila G, Mujeeb-Kazi A, Rajaram S. Inheritance of resistance to *Tilletia indica*(Mitra) in synthetic hexaploids × *Triticum aestivum* crosses. Plant Breeding. 1995;**114:**547-548.

[150] Nielsen J. Inheritance of virulence of loose smut of wheat, *Ustilago tritici*, onthe differential cultivars Renfrew, Florence x Aurore, Kota, and little Club. Canadian Journal of Botany. 1977;**55**:260-263.

[151] Nielsen J. Inheritance of virulence of *Ustilagotritici* on the differential cultivars Carma, red bobs, and a derivative of the cross Thatcher x regent spring wheat. Canadian Journal of Botany. 1982;**60**:1191-1193.

[152] Knox RE, Howes NK. A monoclonal antibody chromosome marker analysisused to locate a loose smut resistance gene in wheat chromosome 6A.Theoretical Applied Genetics. 1994;**89:**787-93.

[153] Knox RE, Campbell HL, Clarke FR, Menzies JG, Popovic Z, ProcunierJD,Clarke JM, DePauw RM, Cuthbert RD, Somers DJ. Quantitative trait loci forresistance in wheat (*Triticum aestivum*) to *Ustilago tritici*. Canadian Journal of Plant Pathology.2014;**36**:187-201

[154] Knox RE, Campbell H, Menzies JG, Popovic Z, Procunier JD, Clarke JM,DePauw RM, Singh AK. Quantitative trait locus for loose smut resistance (*Ustilago tritici*) in wheat (*Triticum aestivum*). Lethbridge: XVI Biennial Workshop on the Smuts and Bunts; 2010

[155] Procunier JD, Knox RE, Bernier AM, Gray MA, Howes NK. DNA markers linkedto a T10 loose smut resistance gene in wheat (*Triticum aestivum* L.).Genome. 1997;**40**:176-179

[156] Kassa MT, Menzies JG, McCartney CA. Mapping of the loose smut resistancegene Ut6 in wheat (*Triticum aestivum* L.). Molecular Breeding. 2014;**33**:569-576

[157] Dhitaphichit P, Jones P, Keane EM. Nuclear and cytoplasmic gene control ofresistance to loose smut (*Ustilagotritici* (Pers.) Rostr.) in wheat (*Triticum aestivum* L.). Theoretical Applied Genetics. 1989;**78**:897-903.

[158] Syukov VV, Porotkin SE. Genetics of common wheat's (*Triticum aestivum* L.) resistance to loose smut (*Ustilagotritici* (Pers.) Jens.) review. Russian JournalGeneticsApplied Research. 2015;**5**:55-9.

[159] Takabayashi N, Tosa Y, Oh H S,Mayama S. A gene-for-gene relationship underlying the species-specific parasitism of *Avena/Triticum* isolates of *Magnaporthe grisea*on wheat cultivars. Phytopathology.2002;**92**: 1182-1188

[160] Zhan S W, Mayama S, Tosa Y. Identification of two genes for resistance to *Triticum* isolates of *Magnaporthe oryzae*in wheat. Genome. 2008;**51**: 216-221

[161] Tagle A G, ChumaI, Tosa Y. Rmg7, a new gene for resistance to Triticum isolates of *Pyriculariaoryzae*identified in tetraploid wheat. Phytopathology. 2015 https://doi.org/10.1094/PHYTO-06-14-0182-R

[162] Anh V L. Rmg8, a new gene for resistance to Triticum isolates of *Pyricularia oryzae* in hexaploid wheat. Phytopathology. 2015;**105**:1568-1572

[163] Anh V L.Rmg8 and Rmg7, wheat genes for resistance to the wheat blast

fungus, recognize the same avirulence gene AVRRmg8. Molecular Plant Pathology.2018;**19**:1252-1256

[164] Wang S.A new resistance gene in combination with Rmg8 confers strong resistance against triticum isolates of *Pyricularia oryzae* in a common wheat landrace. Phytopathology. 2018;**108**: 1299-1306

[165] Song QJ, Shi JR, Singh S, Fickus EW, Costa JM, Lewis J, Gill BS, Ward R, Cregan PB (2005) Development and mapping of microsatellite (SSR) markers in wheat. Theor Appl Genet 110:550-560

[166] Chhuneja P, Kaur S, Garg T, Ghai M, Kaur S, Prashar M, Bains NS, Goel RK, Keller B, Dhaliwal HS, Singh K (2008a) Mapping of adult plant stripe rust resistance genes in diploid a genome wheat species and their transfer to bread wheat. Theor Appl Genet 116:313-324

[167] Chhuneja P, Kaur S, Goel RK, Aghaee-Sarbarzeh M, Prashar M, Dhaliwal HS (2008b) Transfer of leaf rust and stripe rust resistance from Aegilops umbellulata Zhuk. to bread wheat (*Triticum aestivum* L.). Genet Resour Crop Evol 55:849-859

[168] Chhuneja P, Kaur S, Singh K, Dhaliwal HS (2008c) Evaluation of *Aegilops tauschii* (L.) germplasm for Karnal bunt resistance in a screen house with simulated environmental conditions. Plant Genet Resour Charact Util 6:79-84

[169] Malik R, Brown-Guedira GL, Smith CM, Harvey TL, Gill BS. Genetic mapping of wheat curl mite resistance genes Cmc3 and Cmc4 in common wheat. Crop Sci. 2003;43:644-650.

[170] Lanning SP, Fox P, Elser J, Martin JM, Blake NK, Talbert LE. Microsatellite markers associated with a secondary stem solidness locus in wheat. Crop Sci. 2006;46:1701-1793.

[171] Sherman JD, Weaver DK, Hofland ML, Sing SE, Buteler M. Identification of novel QTL for sawfly resistance in wheat. Crop Sci. 2010 50:73-86.

[172] Lapitan NLV, Peng J, Sharma V. A high-density map and PCR markers for Russian wheat aphid resistance gene Dn7 on chromosome 1RS/1BL. Crop Sci. 2007;47:811-820.

[173] Liu XM, Smith CM, Gill BS, Tolmay V. Microsatellite markers linked to six Russian wheat aphid resistance genes in wheat. Theor. Appl. Genet. 2001;102:504-510.

[174] BerzonskyWA, Ding H, Haley SD, Harris MO, Lamb RJ. Breeding wheat for resistance to insects.Plant Breed. Rev. Vol. 22. 2010. DOI: 10.1002/9780470650202.ch5

[175] Zhu LC, Smith CM, Fritz A, Boyko EV, Voothuluru P, Gill BS. Inheritance and molecular mapping of new greenbug resistance genes in wheat germplasms derived from *Aegilops tauschii*. Theor. Appl. Genet.2005;111: 831-837.

[176] Gharalari AH, Fox SL, Smith MAH, Lamb RJ. Oviposition deterrence in spring wheat, Triticum aestivum, against orange wheat blossom midge, *Sitodiplosismosellana*: implications for inheritance of deterrence. Entomol. Exp. Appl.2009;133:74-83.

[177] Thomas MB. Ecological approaches and the development of "truly integrated" pest management. Proc. Natl. Acad. Sci. USA. 1999; 96:5944-5951.

[178] Andersen E J, Ali S, Byamukama E, Yen Y, Nepal M P. Disease resistance mechanisms in plants. Genes. 2018; **9**:339.

[179] Appels R, Eversole K, Feuillet C, Keller B, Rogers J, Stein N.Shifting the

limits in wheat research and breeding using a fully annotatedreference genome. Science. 2018;**361**:eaar7191.

[180] Steuernagel, B.,Witek, K., Krattinger, S. G., Ramirez-Gonzalez, R. H., Schoonbeek, H.-J., Yu, G., et al. (2018). Physical and transcriptional organisation of the bread wheat intracellular immune receptor repertoire. bioRxiv [Preprint].

[181] Lorenz AJ, Chao S, Asoro FG, Heffner EL, Hayashi T, Iwata H, Smith KP, Sorrells ME, Jannink JC. Genomic selection in plant breeding: knowledge and prospects. Advance Agronomy. 2011;**110**:77-123

[182] Rutkoski J, Poland JA, Singh RP, Huerta-Espino J, Bhavani S, Barbier H, Rouse MN, Jannik J-L,Sorrells M. Genomic selection for quantitative adult plant stem rust resistance in wheat.Plant Genome. 2014;**7**(3):1-10

[183] Mirdita V, He S, Zhao Y, Korzun V, Bothe R, Ebmeyer E, Reif JC, Jiang Y. Potential and limitsof whole genome prediction of resistance to *Fusarium* head blight and *Septoria tritici*blotchin a vast central European elite winter wheat population. Theoretical Applied Genetics. 2015;**128**:2471 2481

[184] Milne RJ, Dibley KE, Schnippenkoetter W, Mascher M, Lui ACW, Wang L, Lo C, Ashton AR, Ryan PR, Lagudah ES (2019) The wheat gene from the sugar transport protein 13 family confers multipathogen resistance in barley. Plant Physiol 179:1285

[185] Moore JW, Herrera-Foessel S, Lan C, Schnippenkoetter W, Ayliffe M, Huerta-Espino J, Lillemo M, Viccars L, Milne R, Periyannan S (2015) A recently evolved hexose transporter variant confers resistance to multiple pathogens in wheat. Nat Genet 47:1494-1498

Chapter 3

Insect Pest Complex of Wheat Crop

Mirza Abdul Qayyum, Shafqat Saeed, Unsar Naeem-Ullah, Amar Matloob, Muhammad Wajid, Abou Bakar Siddique, Rumail Shahid, Hafiz Ubaid Ur Rehman Zia, Huda Bilal and Muhammad Ramzan

Abstract

Wheat *Triticum aestivum* L. is grown on broad range of climatic conditions because of edible grains, cereal crop and stable food of about 2 Billion peoples worldwide. Additionally, it is the rich source of carbohydrates (55–60%), vegetable proteins and contributed 50–60% daily dietary requirement in Pakistan. Globally, wheat crops is grown over 90% area of total cultivated area; facing devastating biotic and abiotic factors. The estimated economic losses in wheat quantity and quality are about 4 thousands per tonne per year including physical crop losses and handling. Economic losses of about 80–90 million USD in Pakistan are recorded due to inadequate production and handling losses. Wheat agro-ecosystem of the world colonizes many herbivore insects which are abundant and causing significant losses. The feeding style of the insects made them dispersive from one habitat to another imposing significant crop loss. Areas of maximum wheat production are encountered with either insect which chew the vegetative as well as reproductive part or stem and root feeders. This chapter provides the pest's taxonomic rank, distribution across the globe, biology and damage of chewing and sucking insect pest of wheat. It is very important to study biology of the pest in accordance with crop cycle to forecast which insect stage is economically important, what the proper time to manage pest is and what type of control is necessary to manage crop pest. The chapter will provide management strategies well suited to pest stage and environment.

Keywords: Wheat crop, economic losses, biology, insect pests, management

1. Introduction

Wheat is undoubtedly one of the major cereal crop, staple food and rich portion of daily intake for much of world's population. With annual global production over 770 MT from 220 M hectares, it is a grain of life. The cultivation of wheat started about 10,000 years ago as part of the Neolithic revolution which state a transition from hunting and gathering of food to settle agriculture. Earlier cultivated forms of wheat were diploid (einkorn) and tetraploid (emmer) with known initial origin of the south-eastern part of Turkey. Hexaploid bread wheat that is currently widely adapted in about 95% area of world.

Though wheat was one of pesticide free crop in major areas of the world, however the things are not the same now. Today, all crop production practices are being highly challenged by biotic and abiotic stresses. Biotic stresses especially insect pests and diseases cause devastating damage in terms of yield and quality. On average pests cause 20–37% yield losses worldwide which is translating to approximately $70 billion annually.

Wheat is damaged by sucking and chewingtypes of pests. The list of insect pests damaging different stages of wheat crop varies from region to region, however the complete list of insect pests is around 100. It is therefore important to understand biology of insect pest simultaneously with the crop biology to understand when, where and what chemical should be used to control specific insect/pest more effectively. In this review, we have outlined major insects of wheat along with their biology and control strategies to minimize grain yield losses.

2. Chewing insects' pests of wheat

2.1 Wheat termite *Microtermes obesi* Holm. (Termitidae: Blattodea)

2.2 Taxonomy

Wheat termites belong to order (Earlier Isoptera) Blattodea consisting of 9 families exists worldwide. The families of termites were further classified as monogeneric families including Mastotermitidae (holotype *Masotermes darwinensis* in Australia), Indotermitidae (holotype *Indotermes* in India), Stylotermitidae (holotype *stylotermes* in India), Serrtermitidae (holotype *Serritermes serrifer*in Brazil). The family Termitidae comprises 145 Genera and have 4 subfamilies [1, 2] along with near about 3000 described species [3]. Termite fauna of the subcontinent comprised 337 described species and subspecies of 54 genera. Among these, 16 species found to be damaging to Wheat crops in Asia, of which dominant species are *Odentotermes obesus* and *Microtermes obesi* [4]. However, *Microtermes obesi* is known to be the most important pest of wheat. The taxonomic classification showed that *Microtermes obesi* belongs to order Blattodea and family Termitidae.

2.2.1 Distribution

Termites found all over the world except the Antarctic region. Termites distributed to Tropical, subtropical and temperate regions Worldwide. Termite's diversity is found to be very high in the South American region compared to North America and Europe. Out of 3000 known species of termites are extremely abundant if African region. In Asia the main distribution is restricted to China, India, Pakistan, Sri Lanka and Veitnam totaling 435 species. *Microtermes obesi* commonly known as Wheat termite is a very small species of genus *Microtermesobesi* is restricted to wheat habitats of south and Southeast Asian countries including India, Pakistan, Sri Lanka, Thailand and Vietnam [5, 6].

2.2.2 Biology

Termites undergo a developmental process as in case of other insect species known as incomplete metamorphosis with egg, nymph and adult stages [7]. Nymphs are small entities resembling adults, molts as they grow converted into adult stage. A nymph usually undergoes 3 molts [8]. During the summer months after monsoon, fertile reproductive caste of termites leaves its colony for nuptial flight. After successful fertilization the queen increased in size from 9 to 11 cm and laid around 70,000 kidney shaped eggs that will hatch in nearly 30–90 days. Usually in full reproductive colonies 80–90% individuals belong to the workers caste and 10% Soldier caste [9]. After sometimes they are produced into full adults with wings and reproductive or fertile females which can fly for nuptial flight to repeat cycle for new colony [10].

2.2.3 Damage

Under field conditions, the termites (*M.* obesi) are predominant insect pests causing severe losses in irrigated areas to about 20–40% [11]. In severe conditions, the yield losses might reach up to 40–80% due to this pest [4]. The termites infest wheat crop in Rabi season most of the time soon after sowing to maturity stages [12]. Young workers chew the young and tender part of the wheat plant resulting in dislodgment of seedlings. The mature stage of the wheat plant is also damaged by the workers of the termites causing the plant to dry and produce white ears at earhead stage [13].

2.3 Wheat armyworm *Mythimna separata* Walk. (Noctuidae: Lepidoptera)

2.3.1 Taxonomy

M. separate Walk. commonly known as Oriental armyworm is a minor insect pest of wheat crop. The name was first described by Francis Walker in the 19th century. The taxonomic classification showed that Wheat armyworm (*Mythimna separata* W.) belongs to order lepidoptera and family Noctuidae that noxiously feed on wheat.

Many synonyms of this name were used in the literature are *Pseudaletia separata* Walk. *Cirphis separate* and *Leucania separata.*

2.3.2 Distribution

The wheat armyworm is present in various wheat growing agro-ecosystems from Asia to Australian continent between 45 north and 45 south Latitude and 60 east to far 170 West Longitude. It is found in 27 countries including China, Japan, Pakistan, and India also in Pacific islands from tropics to temperate climatic regions [14].

2.3.3 Biology

The fertile females lay maximum eggs from 500 to 900, spherical and milky white in color to approximately 2000 eggs. The eggs are laid in clusters on or underside the young seedlings or in soil. The eggs will hatch in 2–7 days after that larva emerges and lasts long for 14–22 days. The mature larva possesses green to pink color having greenish to brownish black stripes on the entire body length [15]. Pupation is usually done in soil but can be done under dry leaves or fresh stubbles or fresh tillers as well. The pupae are shiny brownish yellow in color and last upto13 days. Thus, the whole span lasts in about 35–40 days which may repeat multiple times in each year.

2.3.4 Damage

M. separata Walk. damages the yield losses as this is influenced by the environmental conditions of the area and the stage of wheat crop. Innovations in farming systems such as introduction of high yielding cultivars, balanced fertilizer use and crop rotation of wheat increased the chances of high yield loss by this pest [16]. Severe attack must lead to decreased productivity by reduction of quantity as well as quality of grains [17]. Due to its polyphagous nature, it causes severe economic losses in crop production worldwide [18]. The young larvae feed on younger plants and due to its short life cycle results in heavy infestations with much heavier loss to young tillers [19]. In the past, loss due to *M. separata* Walk.was recorded as 10–30% in Wheat crop [20]. With every increase of 1% in leaves consumed by the armyworm decrease the yield by 0.07–0.88 g per plant from booting to panicle stage.

During the young vegetative stage of wheat plant, the damage is more prominent with extensive defoliation. Young larvae may feed at lemma and palea of young grain as well as male part anther of mature flowers. The larvae cut the young seedlings so often the damage is restricted to a single part of the field. During the grain formation stage, the larvae feed upon the panicles from the basal part of the plant causing it to bend downwards and sometimes the plant may fall down.

2.4 Wheat shoot fly *Atherigona naqvi* Steylskal (Muscidae: Diptera)

2.4.1 Taxonomy

The taxonomic classification was described by Steyskal in 1966. In Asia, it was first reported in wheat agro-ecosystem by [21]. However, in the subcontinent, 5 different species from genus *Atherigona* were reported by [22] too. The taxonomy of *A. naqvi* Seyskal. showed that it belongs to order Diptera and family Muscidae renowned for its damage to wheat crop.

2.4.2 Distribution

The Genus *Atherigona* predominantly comprises species of shoot fly which mostly affect maize crop only worldwide. However, this shoot fly species is responsible for serious threats to wheat agro-ecosystems across the globe. Species from genus *Atherigona* are mainly distributed in Pakistan [23], India [24], Thailand [25], and Africa both in East and West African regions [26, 27], and Egypt [28].

2.4.3 Biology

A. naqvi Steyskal adult shoot flies are grayish brown in color having relative smaller size ranging 4–5 mm than common house fly species. Fertile female flies lay eggs on the underside of the tender seedlings and near the base of the stem. Usually 15–25 elongated eggs, cylindrical like boat milky white along with projections with usual longitudinal ridges are laid. Eggs hatched in just 1–3 days after that tiny maggot emerged, starting to creep on to the leaf sheaths of the tillers. Larval period lasts for 7–10 days along with 3–4 larval instars. Pupation usually takes place inside the stem, making the barrel shape a darkish brown puparium. Adults who are free living usually live for approximately 4 days and ultimately complete their short life cycle in about 3–4 weeks.

2.4.4 Damage

The damage is usually done by the immature larvae of all instars. After the emergence of young seedlings, usually 3–4-week-old young seedlings are targeted by the maggots. After hatching, maggots feed the young growing tissues of the plant resulting in drying of central shoot by chewing the central phloem tube produce white dry seedling known as dead hearts. Dead hearts due to dryness can be pulled easily. Sever infestations resulted in bushy appearance of young tillers.

2.5 Surface grasshopper or cutworm *Chrotogonous trachypterus* Blanchard

2.5.1 Taxonomy

The name *Chrotogonous trachypterus* was given by Blanchard in 1836 to tribe *Chrotogonini* Bolivar (1904) and family Pyrgomorphidae. Surface grasshoppers are multivorous, stout, muddy in color, poly phagous insect feeding on almost all the foliage and green tender shoots belonging to order orthoptera and family Pyrgomorphhidae.

Infraspecies: *Chrotogonous (trachypterus) trachypterus.*

2.5.2 Distribution

C. trachypterus is distributed to many countries worldwide. However, Asian countries' including Pakistan, India, Bangladesh and Nepal are known for their maximum abundance. Iran and Afghanistan are also facing a serious problem regarding this pest. Locally it is present in all across Pakistan including Province Punjab, Balouchistan, Sindh, KPK and GB [29].

2.5.3 Biology

C. trachypterus deposits her eggs inside usually 4–5 cm deep in the soil with slight moisture inside. Female digs a hole with the help of an ovipositor and by means of collateral glands; eggs are deposited along with glutinous secretions in a waterproof egg pod. The female covers her tiny yellowish eggs of 7–8 mm by pushing the soil or sand by hind legs [30]. Eggs will hatch in 12–17 days and tiny nymphs of Pale-yellow color which later turned dark brown undergo 5–6 instars. Nymphs are wingless and smaller in size compared to adult. Nymphal period lasts for 13–17 days.

Adults are much larger than nymphs, have well developed mandibles and wings too. Sexual dimorphism is present as a female has four ovipositors to lay eggs and is usually bulky than males whereas males are smaller and rounded [31].

2.5.4 Damage

C. trachypterus is a polyphagous insect and usually present throughout the year. It damages the seedling stage of a variety of crops growing worldwide. Both nymph and adult feed on tillers of wheat plants. Severe attack results in repeated sowing of the crop. Wheat crop is one of its host plants worldwide [32]. The initial development of the seedling is the prime source of grain yield; so, seedling establishment is critical for better productivity. Nymphs and adults feed on young tillers so that in severe attacks the crop failed, and re-sowing had to be done [33]. Among different host plants, wheat seedlings are the most preferred one for *C. trachypterus* [34].

2.6 Pink stem borer *Sesamia inferens* Walker

2.6.1 Taxonomy and nomenclature

Francis Walker in 1856 described *Sesamia inferens* (Noctuidae: Lepidoptera) for the first time. Some common names of the Pink borer are Asian pink stem borer, Graminous stem borer, Pink borer of rice and purple stem borer. Literature has been reported on the synonymy of *S. inferens* and the synonyms are *Leucania inferens* Walker, 1856, *Leucania proscripta* Walker, 1856, Sesamia tranquilaris Butler, 1880, Nonagria innocens Butler, 1881, *Sesamia corticoides* Strand, 1920, *Sesamia kosempoana* Strand, 1920, *Sesamia sokutsuana* Strand, 1920, and *Sesamia hirayamae* Masturmura, 1929.

2.6.2 Worldwide distribution

According to CABI [35, 36], the current distribution of *S. Inferens*is **Asia**, Pakistan, China, India, Japan, Vietnam, Singapore, Korea, Indonesia, Sabah, Taiwan, Hong Kong, Ceylon, Burma and Thailand. Pink Stem Borer of wheat is also reported in **Australian** and **Pacific** islands.

2.6.3 Biology

Life span of *Sesamia inferens* lasts in 40–50 days under favorable conditions [37]. Adults lay globular and creamy white eggs at the base of the wheat plant [38] which may range from 120 to 348 [39]. Egg color changes from creamy white to brown before hatching [40]. Larvae emerge from the fertilized eggs within a week. The newly hatched larvae are pinkish in color with a reddish-brown head [41]. Six instars of larvae take 23–39 days for entering into pupal stage [42]. Full grown larvae measure 30 mm in length [37] and pupate in stem galleries. Adults have straw-colored forewings with a trivial dark brown streak. Males have Pectinate antennae while females have filiform [43]. Under tropical conditions *Sesamia inferens* completes 4–5 generations in a year.

2.6.4 Damage

Pink stem borer *Sesamia inferens* became a key pest in recent time in cereals and can-do considerable yield loss. Different crops of the family Graminae attacked by this polyphagous pest including rice [44], wheat, pearl millet, finger millet and sorghum [45]. Yield losses caused by *S. inferens* in maize may reach from 25.7 to 78.9% [39].

2.7 Shield bug *Eurygaster integriceps*

2.7.1 Taxonomy and nomenclature

Shield bug of wheat is also known as Sunn pest. A total of fourteen species has been reported so far, three of them are considered economically important; *E. integriceps* (Scutelleridae: Hemiptera), *E. mauru* (L.) and *E. austriaca* Schrk. [46].

2.7.2 Worldwide distribution

Sunn pest has a cosmopolitan distribution in **Asia**: Pakistan, India, Afghanistan, Armenia, Azerbaijan, Georgia, Iran, Israel, Jordan, Kazakhstan, Kyrgyzstan, Lebanon, Syria, Tajikistan, Turkey, Turkmenistan, Iraq, Uzbekistan. **Africa**; Algeria. **Europe**; Greece, Cyprus, Macedonia, Bulgaria, Moldova, Romania, Russia, Serbia, Ukraine [35, 36, 47–49].

2.7.3 Biology

The eggs of the *E. Integriceps* are spherical and measure about 1 mm of diameter [48]. At the time of oviposition, eggs look light green, which eventually turns dark as embryo matures [50]. *E. integriceps* has five nymphal instars [51–53]. First instars measure 1.5 mm in diameter and light in color. Nymphs look similar to adults except with 2–3 paired black dots in the midline between the lateral margins of the abdomen. The nymphs cannot be illustrious from closely related spp. and must be identified at adults [53].

Adults are elongated and elliptical, and their color varies from grayish to brown, to red or black [54]. In this concern, Color of the *E. Integriceps* is extremely uneven and has no worth for biosystematics identification [55]. Adult length and width measures 10–12 mm and 6.1–7.1 mm respectively. There is only one generation per year and an obligatory diapause in the adult stage [56–58].

2.7.4 Damage

E. integriceps is a destructive pest of wheat in the Middle East and Central Asia [59, 60]. They do considerable damage to wheat crop from 25% to as much as 100% [61]. Logothetis [62] have reported some severe outbreaks of this pest resulting in complete losses of the crops over large areas. They can destroy all parts of the cereal crops [63]. Nymphs preferably eat young leaves while adults are attracted toward kernels and ears [63]. Symptoms of the damage are shown as a 'deadheart' and withering of the leaves [46].

2.8 Loreyi Leafworm

2.8.1 Taxonomy and nomenclature

Current accepted scientific name ofLoreyi Leafworm *is Leucania loreyi* (Scutelleridae:Hemiptera) although many synonyms has been reported and these are; *Mythimna loreyi* Duponchel, 1827, *Noctua caricis* Treitschke, 1835, *Leucania curvula* Walker, 1856, *Leucania collecta* Walker, 1856, *Leucania exterior* Walker, 1856, *Leucania thoracica* Walker, 1856, *Leucania designata* Walker, 1856, *Leucania denotata* Walker, 1856, *Acantho leucania loreyi* Duponchel, *Cirphis loreyi* Duponchel, *Hyphilare loreyi* Duponchel.

2.8.2 Worldwide distribution

Leucania loreyi has been reported in various subcontinents of the world including **Asia**: Pakistan [64], Afghanistan [65], Lebanon, Azerbaijan [35, 36], Georgia, Iran, Israel [47], Jordan, Kazakhstan, Lebanon, Syria, Tajikistan, Turkey, Turkmenistan, Iraq, Uzbekistan [64]. **Africa**; Algeria. **Europe**; Greece, Cyprus, Macedonia, Bulgaria [66], Moldova, Romania, Russia, Western Siberia [67], and Ukraine [35, 36].

2.8.3 Biology

Leucania loreyi predominantly lay eggs in masses of 2–127 eggs but may lay singly at the last part of the oviposition time [68]. Eggs are laid on the leaf-sheath of the plants of Graminae. Eggs measure 0.5 mm diameter and are discoid in shape. Color of the freshly laid eggs is ashen-yellow or cream colored. Number of eggs laid decreases with increase in temperature above 30°C [69]. Mating starts after two days of emergence [69]. *L. loreyi* has six larval instars at 29' *C* and 70% relative humidity [68]. First instar larvae are transparent, light green and have elongated body. The color of the second and third instar larvae alters to pale green, dark-green respectively. Fourth and fifth instar larvae have the same colouration except having two lateral pale-brown lines. Newly pupated larvae are of yellowish-cream colored, which turns shiny brownish as pupae matures. Larvae pupate three inches below soil surface and measure 17-23 mm long and 5 mm wide [70]. Adults measure 17 mm in length 35-38 mm along their wingspan. Female moths are comparatively larger in size than males. Head and thorax covered with brownish-yellow scales. Forewings are of a rusty-brown color [68].

2.8.4 Damage

Loreyi leafwormis a major pest of graminous crops including wheat and maize [71]. Up till now, *L. loreyi* is not a major pest in Pakistan, but it has been reported to repeatedly cause rigorous damage in China [72] and other Asian countries. *L. loreyi* has been also known to cause damage in rice [73–78].

2.9 Black cutworm *Agrotis ipsilon*

2.9.1 Taxonomy and nomenclature

Ochsenheimer proposed the genus *Agrotis* in 1816. There are a number of synonyms proposed for *Agrotis ipsilon*, some of these are *Agrotis aureolum* Schaus, 1898, *Agrotis bipars* Walker, 1857, *Agrotis frivola* Wallengren, 1860, *Agrotis pepoli* Bertolini, 1974, *Agrotis suffusa* (Shiffermiller), *Agrotis telifera* Donzel, 1837, *Bombyx idonea* Cramer, 1780, *Bombyx spinula*Esper, 1786, *Noctua amenituma* Walker, 1865, *Noctua suffusa* Denis & Schffermuller, 1775, *Phalaena ipsilon* Hufnagel, 1766, *Phalaena ypsilon* (Cramer), *Phalaena ypsilon* Rottenberg, 1776, *Rhyacia pernigrata* Warren, 1912 and *Scotia ypsilon* (Hufnagel).

2.9.2 Worldwide distribution

Black cutworm is also scattered worldwide and causes a huge loss of crop yield. They are distributed in **Asia**; Pakistan, Afghanistan, Armenia, Azerbaijan, Bangladesh, Cambodia, China, India, Israel, Iran, Iraq, Japan, Jordan, Lebanon, Malaysia, Myanmar, North Korea, Philippines, Saudi Arabia, Singapore, South Korea, Sri Lanka, Syria, Turkey, Thailand, Taiwan, United Arab Emirates and Vietnam. **Africa**: Benin, Burkina Faso, Congo, Egypt, Kenya, Liberia, Libya, Madagascar, Malawi, Mali, Mauritius, Morocco, Reunion, Saint Helena, Senegal, South Africa, Sudan, Togo, Tunisia and Zimbabwe. **Europe**; Albania, Austria, Belgium, Bulgaria, Croatia, Cyprus, Yugoslavia, Denmark, Estonia, Finland, France, Germany, Greece, Hungary, Iceland, Ireland, Italy, Norway, Poland, Russia, Spain and the United Kingdom. **North America**; Canada, Dominican Republic, Honduras, Mexico and the United States. **South America**; Argentina, Bolivia, Brazil, Chile Colombia and Piru [35, 36, 79–81].

2.9.3 Biology

Agrotis ipsilon completes its life cycle in 50–77 days at 22'C-25'C temperature and 69–77% relative humidity [82]. Adult female moths lay 145–200 eggs. Egg laying goes on peak 5–6 days after mating [82]. They are 0.5 mm long and 0.45 mm high, semi-circular and have erect and parallel patterned terns. Recently laid eggs have milky color which turns grayish black at the time of hatching [83]. There are six instars of larvae of *A. ipsilon* [82]. Larvae color is grayish-black while ventral and sub-ventral abdominal sides are lighter in color.

Third to sixth instar larvae are 7 mm, 10-20 mm, 20–30 and 37–47 mm long [35, 36, 83]. Larvae of *A. ipsilon* can be distinguished from others with having these diagnostic characters; Stigerous D_2 tubercle large, dorsal anterior tubercle almost one third as long as posterior tubercle, pigmented black spiracles with convex granules [35, 36]. Papae are brownish in color which turns blackish at the time of adult emergence and measures 17–25 mm. Fifth and sixth abdominal tergites have distinct punctures on pleural sides; hook-like spines are present at the apex of the sixth abdominal segment [83]. Adults of the black cutworm are 16–23 mm long, spreaded wingspan measures 35–54-mm [35, 36]. There are three sections in the forewing internal transverse line, external transverse line and some distinct spots: clavate spot, reniform spot, ring spot and sword spot. Hindwings are grayish in color and do not have any markings. Males have feathers type antennae while females have filiform [35, 36].

2.9.4 Damage

Black cutworm has a broad has range and feeds on all crops and pasture plants. Newly emerged crop seedlings are attacked by the full-grown larvae migrated from summer and autumn weeds. Species of the *Agrotis* are the worst destroyers that they may attack on the whole fields of cereal crops. Early instars of larvae feed on the tender tissues of the foliage, but as they grow, they suppose their classic cutworm 'felling' activity [84].

3. Management of chewing insect pests of wheat

Seed is the basic building block of a crop. As seed is healthy insect and disease free more the yield is obtained. Insects are small creatures but are highly reproductive in nature so that they compete with humans for resources. Controlling of insects at appropriate time is the key point to obtained higher and healthy yields.

3.1 Seed treatment

Seed treatment is one of most important control measures which reduce the chances of insect's pests attack because eggs of some insects are glued to seeds and spores of different seed born diseases are attached so that seed treatment is necessary to control the insect's population at its initial stages. Different Insecticides were used to treat seeds against different insects' pests. Fall army worm *Spodoptera frugiperda*, green bug a, hessein fly often present in wheat crop in numbers so that they have a great impact to limit the yield of crop. In this case different measures can be used to limit these pests' infestations. Systemic seed treatment is which is an effective method to control these pests' infestations. Insecticides like imidacloprid treatment of seeds reduce infestation of Russian wheat aphid 27–85 days in wheat and barley crops [128, 139, 140].

3.2 Quarantine

A quarantine pest is "a pest of potential economic importance to that piece of land where it is present but not widely distributed and controlled or the endangered zone where pest does not presentyet. Chewing insect pests requires quarantine measures. Different chewing pests which were not reported or not widely distributed to wheat zone of Pakistan. Arecent exampleof it is fall armyworm that were 1st reported in Nigeria in west Africa in 2016 and within a short duration of time it was reported in 44 countries of Africa. In 2018 it was 1st reported in India [85, 86] that moved to Bangladesh, China, Sri lanka, Thailand, Myanmar [87]. Suitable and perfect environmental conditions for fall armyworm reproduction and wide range of host plants availability in Pakistan. corn and wheat zones are endangered, and several articles were published in newspapers. The international maize and wheat improvement center (CIMMYT) have cautioned Pakistan to make efforts against fall armyworm a potential threat to maize and wheat in Pakistan.

World Trade Organization agreement on the application of sanitary and phytosanitary measures and the international plant protection convention of the food and agriculture organization of united nations and convention for biological diversity all these organizations highly recommended that prevention is the most effective control of invasive species within a minimum cost. It shows better and cost-effective results where it is adopted. The IPCC released a summary of international standards and phytosanitary measureswhich includes all sanctions and guidelines for whole trade processes. Economically harmful species of plants and plants product are black-listed and banned from entering in the whole continent in Europe. The most cost-effective control against invasive species of insect pests is to inspect the incoming consignments with sanitary and phytosanitary inspections at borders this is the last weapon of defense which can be used against invasive species otherwise their control is very difficult.

3.3 Biological control

Biological control plays an important role in wheat crop pest management. Negligible use of insecticidesprovides conducive environment for biological agents to flourish and reproduce. Biological control is the most effective control method when using with other compatible controlling techniques for example in IPM we use biological control as an effective component of crop pest management with other controlling practiceslike cultural control and planting resistive varieties against insects' pests. On the other hand, we use selective insecticides against pests when other controlling method failsdue to some biotic or A-biotic factors to keep the population below ETL.

3.3.1 Biological control managers

Biological control managers of insects are divided intopredators, parasites and pathogens.

3.3.1.1 Predators

Predators are lions they kill and eat their host within few minutes for example Convergent lady beetles.

3.3.1.2 Parasites

Parasites are internal and external or attack to specific life stages of pests. The most important parasites belong to parasitic Hymenoptera puncture the parasites eggs with their sharp ovipositor and lay single egg eggs hatched in 6 to 7 days and larvae feeds on these eggs. Some parasitoids lay eggs directly larvae after hatching parasitoid larvae feed on internal parts of parasite and emerge dead larvae and their mummies left and found in the fields.

3.4 Physical control

physical control is also possible in some insects like larvae of some insects identified and picked from plants individually it is also done with modified method by **rope dragging** in wheat fields against different chewing insects' pests like army worm larvae feed on wheat fields and aphids and some other insects when.

3.5 MST and RIDL

Release of sterile male to reduce the population of an insect pest is a molecular approach and it is also practically performed against different pests like lepidopterans and dipterans. Male sterile techniques are used against chewing pests of wheat is a good approach to reduce the population without affecting our biological fauna.

RIDL is defined as release of insect with dominant lethal gene this technique was used against different pests like fruit flies and this control is also use against different dipterans. RIDL approaches as an insect having a dominant gene surviveand cause lethality in conditions when mating with a female. The survivors refer to a carrier of set of genes and strategies having bisex lethal, flightless females and non-sex specific late-acting lethal systems.

3.6 Legislative control

Wheat legislative control adds as timely sowing of wheat crop with good practices, recommended density of plants,

3.7 Push pull strategies

A behavior manipulation strategy known as Push Pull technique, which is widely used against different insect pests. The term Push pull coined by Australians to control the pests without use of hazardous insecticides. This strategy used against different pests to reduce their abundance. Australians use this strategy against different Helicoverpa species in cotton. Push pull technique combining with other control methods like natural enemies gives good results [88].

3.8 GMOs

Genetically modified organisms are used to kill insect pests and also genetically modified verities of different crops used to control the pests without using of any chemicals. Bt corn and Bt cotton is one of most popular strategies which are currently use against pests [89].

4. Sucking insect pests of wheat

4.1 Chinch bug *Blissus leucopterus*

4.1.1 Taxonomy

Chinch word in Spanish means Pest. The family lygaeidae genus Blissus which contain chinch bug species, yet the taxonomy of the genus is poorly understood. Chinch bug, though native of tropical America but extended its range to the world. It subdivided into two species *Blissus leucopterus* and subspecies *B. leucopterus leucopterus* which are known as Chinch bug and hairy chinch bug respectively. Before and in 1831, the original species combination was Lygaeus leucopterus Say [90] The genus name was replaced in 1835 burmeister. The species name combination of two word leuco which means lack of color and pterus means wings [91]. However, it belongs to order hemiptera (suborder: Heteroptera) and family Lygaeidae. Sub species are *B. leucopterus leucopterus* (Say), *B. leucopterus hirtus* (montandan).

4.1.2 Distribution

Chinch bug, *B. leucopterus* (say) native to the new world, found throughout Americas south as well as the North America region. Chinch bug spread Virginia to Georgia extending to south Dakota and Texas in east and west respectively [92].

4.1.3 Biology

B. leucopterus passes two generations per year, a complete life cycle occurs in 30–60 days. Eggs are elongate-oval and rounded at one end, truncate shape at the other end. The eggs are whitish on the first days, turn yellowish after a few days and become red at time of hatching. Female of chinch bug lays eggs in short rows on root, stem, leaf sheath or on soil near the plant [92]. Eggs rate laying by females are 15–20 per day over 2–3 week, producing upto 500 eggs a single female. Eggs are hatching 16 days at 27°C and 8 days at 31°C [93].

There are five nymphal instars with 5, 6, 5, 4, and 6 intervals during each instar when reared at 29°C, under field conditions, the development time may be extended. The normal development time is 30–40 days in normal and may be extended in 60 days. Identification of nymph in early stages, head and thorax are brown, legs are yellowish. These colors are darker as the mature nymph, so the mature nymphs are blackish in color. There are yellowish and whitish colors on the first two segments of abdomen. Wing pads become visible in 3rd instars. Nymphs prefer sheltered locations to feed and aggregate on the stem near the main stem of the plant [92].

Adults are blackish in color; wings nearly attain the end of abdomen and are white in color with blackish spots found near the center. Measurements of adults are 3.5 to 4.5 mm in length.

4.1.4 Damage

Host of chinch bugs consist solely of family gramineae, but also include other grasses and plants. Chinch bug is a plant feeding insect, causing reddish color at the site of feeding and death of the plant. Plant growth can be stunted, or dead by a large number fed on plants. The losses by chinch bug were estimated at 19 million dollars in 1989 [94].

4.2 Wheat aphid

4.2.1 Taxonomy

Aphids evolved in Carboniferous period about 280–300 million year ago [95, 96] Many species of aphid attack on wheat crop, three major aphids pest are *Myzus persicae* (Sulz.), *Sitobion avenae* (F.) and *Schizaphis graminum* (Rondani). Sulzer was the first who described *Myzus persicae* in 1776 as *Aphis persicae.* Species have many synonyms which are listed by Borner [97]. The Aphid genus of Myzus belongs to the largest tribe of aphid Macrosiphini [98], which contains fifty five species. *M. persicae* make a species complex which describes as a separate species *M. nicotanae*.

Order: Hemiptera. Suborder: Sternorrhyncha, Family: Aphididae, Genus: Myzus, Sitobion and Schizaphis, Species: *M. persicae, S. avenae and S. graminum.*

4.2.2 Distribution

Myzus persicae (Sulzer), green peach aphid is found throughout the world. The green peach aphid probably belongs to Asian origin, *Myzus persicae* growth is not acceptable when temperature and humidity are not bearable for aphids [99].

4.2.3 Biology

Aphids biology is quite complicated than other insects. In single colony, aphid adults are present in wing form or wingless form. Aphids characters in life cycle is continuous asexual production and of larvae by live birth, parthenogenesis and viviparity respectively. In parthenogenesis, embryos arise from cells without reduction of chromosomal and individuals, so all females are genetically identical to their parents. Sexual reproduction occurs in autumn, female of autumn season oviparous results in the production of overwintering eggs. In the following growth season of plant eggs hatch and produce a series of parthenogenetic generations [100, 101].

4.2.4 Damage

Aphids damage in stages of adults and nymphs. Nymph and adult suck the cell sap of the plant part [102]. Aphids can attain very high populations on young plant tissue, wilting and reducing the growth rate of plants. Losses due to aphid upto one third crops yield [103].

4.3 Green stink bug

4.3.1 Taxonomy

The genus *Nezara* proposed by Amyot and Serville in a group 'Rhaphigastrides' with other spices in 1843. In the family Pentatomidae, Kirkaldy (1909) recognized six subgenera in *Nezara*, now all of which are considered as genera. *Nezara viridula* have color variability and wide distribution in the world, resulting in the form of synonyms [104]. Most existing forms of species are two G-type and O-type, *Nezara viridula var. Smaragdula* is G-type with complete green color and *Nezara viridula* var. *torquata* is O-type with green body and anterior yellowish coloration [105]. It belongs to Order: Hemiptera, Suborder: Hetroptera and Family: Pentatomidae.

4.3.2 Distribution

*Nezara viridula*is referred to as worldwide or cosmopolitan distributed as the species and occurs throughout regions (Tropical, Subtropical and Temperate) Also in Australia [106, 107]. The species *Nezara viridula* expands its range constantly, both in Northern and Southern spheres, by natural dispersal and human translocation [108].

4.3.3 Biology

The development of life stages of this species has been described by Jones in 1918. Females lay eggs in clusters; each cluster contains 60–90 eggs [109]. Fresh eggs are cream in color, and become dark after one day, eggs hatch in 3 days [110]. First instars of *Nezara viridula* are red in color and turn dark by the stadium on the second day [111]. Second to fourth instars are green in color, fourth and fifth instars may be green/dark [112]. Development of insect adults from eggs is approximately 30 days but varies on time period. Female adults start mating in 5 days and male take 6 days for mating. Female deposited the egg within 7 to 8 days after mating. The diapausing of species occurs in adult form and insect diapauses before mating.

4.3.4 Damage

Nezara viridula species is a highly polyphagous insect, attacking both monocot and dicots. Range of plants is 145 plant species which belong to 32 families as host plants [113]. Different generations of species breed utilize and feed different plant species during vegetative stages of host plant in a season. Damage come from feeding of nymphs on podm fruits and seed which results in yield reductions and other aspects like quality and germination of seeds.

4.4 Haplothrips

4.4.1 Taxonomy and nomenclature

Haplothrips was first described by Amyot & Serville in 1843. Synonyms of the *Haplothrips ganglbaueri* Schmutz, 19136 are *Haplothrips ganglbaueri* Schmutz, 1913, *Haplothrips angustus* Hood, 1919, *Haplothrips vernoniae* Priesner, 1921, *Zygothrips andhra* Ramakrishna, 1928, *Haplothrips priesnerianus* Bagnall, 1933, *Haplothrips themedae* Priesner, 1933 and *Haplothrips tolerabilis* Priesner, 1936 [114]. Order: Thysanoptera and Family: Phlaeothripidae.

4.4.2 Distribution

Haplothrips is widely distributed in **Old world**, Pakistan, China, Iran, Japan, Sri Lanka, India, Indonesia, Egypt. **New world**; Central and South America, Australia and New Zealand [115–117].

4.4.3 Biology

Eggs of Haplothrips are cylindrical, rounded from posterior, tapered anterior end, which looks like a knoblike process. Eggs measure 433-500 μ length and 137-150 μ width. Nymphs at the time of egg hatching are microscopic, transparent and amber in color. Color changes from amber to pink after considerable feeding. The length of the first instar before eclosion measures 1100 μ in length. The color of the second instar nymph is glowing red except for the appendages which are dark brownish to black. Second instar mature nymph measures about the same length as that of the adult. The pre pupal stage of the *Haplothrips ganglbaueri* ischaracterized by small antennal sheath, the glassy colored appendages and the lack of wing sheaths. Overall pre-pupal stage is pale red. Adult color is pale red. They have a transparent head, with a dorsal blotch. An occasional adult in the field may be dark red. Length of the adult's measure 1415-2268 μ. Morphology of abdomen is compressed and pointed toward its apex, fringed with setae. Head, thorax, and abdomen lack bristles. Seven to nine accessory cilia present on the wings [118].

4.4.4 Damage

Polyphagous pest *Haplothrips ganglbaueri* severely damage graminous crops such as *Oryza sativa*, wheat *Triticum vulgare*, and *Sorghum vulgare*. It has been known for doing damage to fruiting parts such as inflorescence. Both adults and nymphs preferably feed on inflorescence. Uneven oval and subtle brown patches on the lemma, palea and ovarian tissues of rice were found by Ananthakrishnan and Thangavelu [115].

5. Management of sucking insect pest of wheat

5.1 Cultural methods

Cultural control comprises the modification of regular farm operations that destroy the insects or prevent them from causing injury. This control is to adjust the time of sowing, plowing, irrigation, harvesting and improved farm management. The opinion regarding aphids shows that it damages the wheat badly that is sown earlier and if the cool weather remains until March [119]. Tabasum et al. [120] reported that the crop sown earlier was least affected and the wheat crop can be set aside by doing modification in sowing dates. The early sowing of the wheat crop is the best way to minimize the risk of aphid attack [121]. The abundance of *Coccinella septempunctata* was greater on late-planted wheat than the crop sown earlier [122]. Preferably wheat in Multan should be planted in the last week of November to avoid heavy aphid attack [122].

Intercropping with different crops can increase the natural enemy population in a wheat field for many reasons. The intercrop plants may release chemicals to attract natural enemies and their early establishment in the field. Intercropping with non-host plants seemed to be favorable for the parasitoid's population [123]. The rye-grass strips in wheat fields and wheat– oilseed rape intercropping is used to enhance the number of natural enemies. The population density of ladybeetle and ratio of ladybeetle to *S. avenae* was greater than in the wheat-oilseed rape intercropping field. It is recommended that *Bactris campestris* intercropping with wheat should be encouraged among farmers to maximize the wheat crop profit by reducing the aphid population [124].

5.2 Seed treatment

Seed treatment is environment friendly and economical with excellent control. The systematic and relatively low rate of the application makes it user-friendly for seed dressing and it protects from sucking insect pests by eliminating the repeated needs for sprays. Seed treatment by using neonicotinoid insecticides against piercing-sucking insects, such as aphids is very effective. Ahmed et al. [125] reported that when imidacloprid in combination with tebuconazole is used as seed treatment against *Schizaphis graminum* can control aphids for 8 weeks. The mixture of imidacloprid + Tebuconazle with a relatively low rate of application can make it environment friendly and an effective option for seed dressing against *Schizaphis graminum* [126]. It can be concluded that Actara® and Hombre® as seed treatment could be efficiently utilized for controlling *Sitobion avenae* [127].

5.3 Biological control methods

Natural Biological control is the action of predators, parasitoids, pathogens and plant extracts in maintaining pest density. The natural enemies may help to reduce the sucking pest population from reaching the economic injury level in the Wheat. The aphid parasitoids in Pakistan have been reported by [128, 129]. In Pakistan, *Aphidius* sp. has been recorded parasitizing *S. graminum* attacking wheat crop [130].

A. ervi and *A. colemani* are reported by against wheat aphid [131], while *Diaeretiella rapae* reported by [126]. On wheat aphid parasitism rates started low, as the season progressed, the mean rate of parasitism increased [132].

In wheat, sucking pest populations are effectively restricted by adults and larvae of ladybird beetles, lacewing larvae and larvae of hoverflies. Predators are the parasitoids due to their broader host range and can feed on both egg and larvae stage of pests and also [129]. Coccinellids are the most abundant predator on wheat and cotton for the controlling of the aphid population [133]. *Coccinella septempunctata* is one of the most efficient predators of immature and adult aphids on wheat [134]. As the biological control agents, syrphid flies against *S. graminum* may provide a complementary management method [135]. One of the voracious predators of all the aphids exposed eggs and small nymphs are *Chrysoperla carnea* [136].

There are several botanicals derived from plant oils extracted from leaf and seeds have been used to control aphids in Pakistan. *Moringa oleifera* and *Eucalyptus oblique* leaves showed higher mortality of *S. avenae* [137]. *Azadirachta indica* seed kernel extracts to control *S. avenae* are effective as imidacloprid [138]. Abid [139] concluded that tobacco caused maximum mortality against all instars of *S. graminum* and *S. avenae* followed by neem, dhatura and onion. Iqbal et al. [140] treated the aphid by different botanicals, Orange Peel extract exhibited the maximum mortality of aphid followed by Garlic and Tobacco. *Azadirachta indica* and the entomopathogenic fungi *Beauveria bassiana* or *M. anisopliae* exhibited efficacy against *S. avenae* [126].

5.4 Host plant resistance

Plants can resist invading insects and diseases. The plants with this ability can be attributed to their morphological and chemical characteristics. Moreover, resistant plants can change their physiology in case of invasion and compensate for the damage caused by the pests. Planting resistant cultivars is a simple and effective method to reduce its damage. Shahzad et al. [121] proved that Galaxy 2013 gives higher yield and can tolerate aphid damage. Wheat genotypes can play a vital role to suppress the aphid population, Sarsabz, Kiran-96 and Khirman varieties were shown to be resistant [141]. Results proved that 6309–2103 shows resistance among other varieties and has the lowest Aphids population density [142]. Shafaq-06 is more susceptible and 9114 is relatively more resistant wheat varieties lines against the aphid population [134].

Author details

Mirza Abdul Qayyum[1*], Shafqat Saeed[1], Unsar Naeem-Ullah[1], Amar Matloob[2], Muhammad Wajid[1], Abou Bakar Siddique[1], Rumail Shahid[1], Hafiz Ubaid Ur Rehman Zia[1], Huda Bilal[1] and Muhammad Ramzan[1]

1 Institute of Plant Protection, Muhammad Nawaz Shareef University of Agriculture, Multan, Pakistan

2 Department of Agronomy, Muhammad Nawaz Shareef University of Agriculture, Multan, Pakistan

*Address all correspondence to: qayyum.mirza@mnsuam.edu.pk

References

[1] Roonwel, M. L., & Chhotani, O. B. (1989). *Fauna of India*. Esoptera (Termites).

[2] S. Khalid, S. Ali, K. Akhtar, M.S. Khan, A. Ali, A. Samad, S. Hussain, S. Fahad and F. Khan. 2015. Preferential influence of wheat genotypes on the distribution pattern and population dynamics of cereal aphids and their natural enemies in Peshawar valley. Pakistan J. Zool 47:223-233.

[3] Jan Šobotník and Cecilia AL Dahlsjö (2017). Isoptera. Reference Module in Life Sciences doi:10.1016/B978-0-12-809633-8.02256-1

[4] Chhillar B S, Saini R K and Roshanlal K (2006) Emerging Trends in Economic Entomology. Publ. by *CCSHAU Press, Hissar*, 191-192.

[5] Akhtar MS (1975) Taxonomy and zoogeography of the termites (Isoptera) of Bangladesh. Bull Dept Zool Univ Panjab (N B) 7:1-199

[6] Karunaratne, W. (2012). *An annotated checklist of termites (isoptera) of Sri lanka* (Doctoral dissertation, Department of Botany, Faculty of Science, University of Peradeniya).

[7] Davis, P. "Termite Identification". (http://webarchive.loc.gov/all/20090612011547/http://agspsrv34.agric.wa.gov.au/ento/termites.htm). Entomology at Western Australian Department of Agriculture. Archived from the original (http://agspsrv34.agric.wa.gov.au/ento/termites.htm) On 2009-06-12.

[8] Neoh, K.B.; Lee, C.Y. (2011). "Developmental stages and caste composition of a mature andincipient colony of the drywood termite, *Cryptotermes dudleyi* (Isoptera: Kalotermitidae)".Journal of Economic Entomology. 104 (2): 622-628. doi:10.1603/ec10346.

[9] Kaushal Kishor, Jitendra Kumar, Vikrant, S.K. Dotasara and Sanjeev Sharma. 2017. Wheat Crop Damaged by Termite and Co-Related of Different Doses of Insecticidal Treatment Compared with Yield in Standing Crop. *Int.J.Curr.Microbiol.App.Sci*. 6(12): 449-454. doi: https://doi.org/10.20546/ijcmas.2017.612.055

[10] NIPHM (2014) AESA based IPM package wheat. National Institute of Plant Health Management, Hyderabad, pp. 1-72

[11] Mishra R.D. Sharma, R.K., Singh, K.P., Parsohan, P.A. Tiwari, A.N., Verna, R.S. and Jaiswal, (2003). Wheat Research Pantnagar, Research Bulletin No. 132, Directorate of Experiment Station, GBPUA&T, Pantnagar. Uttranchal: 47-49.

[12] Sharma AK, Sahaan MS, Babu KS (2009) Wheat crop health. Newsletter 14(4):23-27

[13] Pardeshi MK, Kumar D, Bhattacharyya AK (2010) Termite (Insecta: Isoptera) fauna of some agricultural crops of Vadodara, Gujarat (India). Rec Zool Surv India 110(1):47-59

[14] Sharma, H. C., & Davies, J. C. (1983). The oriental armyworm, Mythimna separata (Wlk.). Distribution, biology and control: a literature review. *The oriental armyworm, Mythimna separata (Wlk.). Distribution, biology and control: a literature review.*, (59).

[15] Tanwar, R. K. (2010). Rice swarming caterpillar (*Spodoptera mauritia*) and its management strategies. *Technical bulletin (Ntational Centre for Integrated Pest Management (India))*.

[16] Sharma, H. C., sullivan, D. J. and Bhatnagar, V.S., 2002. Population

dynamics and natural mortality factors of the Oriental armyworm, *Mythimna separate* (Lepidoptera : Noctuidae) in South-Central India. Crop Prot., 21: 721-732.

[17] Khattak, M. A., Riazuddin and Annayatullah, M., 2007. Population dynamics of aphids (Aphididae: Homoptera) on different wheat cultivars and response of cultivars to aphids in respect of yield and yield related parameters. *Pakistan J. Zool.,* 39: 109-115

[18] Jiang, X., Zhang, L., Yang, H., Sappington, T. W., Cheng, Y., & zhi Luo, L. (2016). Biocontrol of the oriental armyworm, Mythimna separata, by the tachinid fly Exorista civilis is synergized by Cry1Ab protoxin. Scientific reports, *6*(1), 1-8.

[19] Bai, W. H., Liu, A. P., Song, Y. F., & Xu, S. T. (1990). Investigation of destructive insects on major forage plants in north China. Grassland of China, (5), 58-60.

[20] Chaudhary, J. P., & Singh, Z. (1980). Outbreak of armyworm, Mythimna separata (Walker) on paddy crop in Haryana [rice, India]. Note. *Haryana Agricultural University Journal of Research*.

[21] Pont, A. C. 1972. A review of the oriental species of *Atherigona Rondani* (Diptera, Cumcidae) of economic importance, p. 27-104. M. G. Jotwani and W. R. Young (eds.) Control of sorghum shoot fly. Oxford and IBM, New Delhi, India.

[22] Panwar, V. P. S., & Sarup, P. (1985). Distribution and host-plants of shoot fly species attacking maize in different parts of the world. Journal of entomological research.

[23] Moiz, S. A., & Naqvi, K. M. (1968). Studies on sorghum stem fly Atherigona varia var. soccata Rondani (Anthomyiidae: Diptera). *Agric.* Pak, *19*, 161-164.

[24] Veda Moorthy, G., V. V. Thobbi, B. H. Matai, and W. R. Young. 1965, Preliminary studies with seed and seed-furrow applications of insecticides for the control of the sorghum stem maggot, *Atherigona indica* Mailoch (Anthomyiidae). Indian J. Agric. Sci. 35: 14-28.

[25] Harwood, R. R., Granados, Y. R., Jamornman, S., & Granados, R. G. (1972). Breeding for resistance to sorghum shoot fly in Thailand. Control of Sorghum Shoot Flv, 208-217.

[26] Barry, D. E. A. N. (1972). Life history and other biological notes on sorghum shoot fly in East Africa. Control of Sorghum Shoot Fly, 119-125.

[27] Langham, R. M. (1968). Inheritance and nature of shootfly resistance. *Sc. Thesig. Ahmadu Bello University, Zaria, Nigeria*.

[28] El Abdin, A. M. Zein. 1980. Review of shootfly research in Sudan, p. 45-46. In Proc. International Study Workshop on the Sorghum Shootfly. International Centre of Insect Physiology and Ecology, Nairobi, Kenya.

[29] Samiullah Soomro and Riffat Sultana. 2020. "Diversity with position of habitat of Pyrgomorphidae Brunner von Wattenwyl, 1874 (Orthoptera: Caelifera) from Khairpur, Sindh", International Journal of Current Research, 12, (07), 12647-12650

[30] Meena S. Bioecology and management of *Chrotogonus trachypterus* Blanchard (Orthoptera; Acrididae), a polyphagus pest of semi arid zone of Rajasthan, 2012; University of Rajasthan, PhD thesis.

[31] Meena S and Singh NP. Applicability of Dyar's law for instars identification of *Chrotogonus trachypterus* Blanchard (Orthoptera: Acrididae), J. Exp. Zool. India, 2009; Vol. 12(1):203-206.

[32] Meena S and Singh NP. Studies on the consumption and utilization of different food plants by *Chrotogonust rachypterus* Blanchard (Orthoptera: Acrididae). Entomon,2010; 35(2): 135-138.

[33] Akhtar, M. 1971. Laboratory feeding tests with *Chrotogonus trachypterus* Blanchard (Orthoptera: Acrididae). Pakistan J. Zoo. 3: 163-167.

[34] Asad, R., M.S. Awan., G.H. Abro and A.A. Shah. 2001. Studies on feeding, copulation, oviposition and defence behaviour of Chrotogonus trachypterus (blanch) Orthoptera: Pyrgomorphidae) under laboratory conditions. Pak. J. Zool. 33 (2): 85-91.

[35] CABI, 2020a. *Agrotis ipsilon* (Black cutworm): Datasheet. Wallingford, UK.https://www.cabi.org/isc/datasheet/3801#REF-DDB-105170.

[36] CABI. 2020b. *Eurygaster integriceps*. Crop Protection Compendium, Wallingford, UK, www.cabi.org/cpc/

[37] Prasad, G. S., & Babu, K. S. (2016). Insect Pest Resistance in Pearl Millet and Small Millets. In Biotic Stress Resistance in Millets (pp. 147-169). Academic Press.

[38] Singh B. Incidence of the pink noctuid borer *Sesamia inferens* (Walker), on wheat under two tillage conditions and three sowing dates in north- western plains of *India.* Journal of Entomology. 2012; 9(6):368-374.

[39] Baladhiya, H. C., Sisodiya, D. B., & Pathan, N. P. (2018). A review on pink stem borer, *Sesamia inferens* Walker: A threat to cereals. Journal of Entomology and Zoology Studies, 6(3), 1235-1239.

[40] Aggarwal R, Singh J, Shukla K. K. 2004. Biology of pink stem borer, *Sesamia inferens* Walker on rice crop. Indian Journal of Ecology; 31:66-67.

[41] Sharma, H., Jaglan, M. S., & Yadav, S. S. 2017. Biology of pink stem borer, Sesamia inferens (Walker) on maize, Zea mays. Journal of Applied and Natural Science, *9*(4), 1994-2003.

[42] Nagarjuna, B., Manjunath, M., & Latha, M. 2015. Biology of maize stem borer, Sesamia inferens (Walker) Noctuidae: Lepidoptera. Journal of eco-friendly Agriculture, *10*(1), 90-91.

[43] Viswajyothi K. 2011. Biology of *Sesamia inferens* Walker on maize vis-à-vis impact of selected environmental variables. M.Sc. thesis submitted to the Punjab Agricultural University, Ludhiana.

[44] Khan Z.R., Litsinger J.A., Barrion A.T., Villanueva F.F.D., Fernandez N.J. and Taylor L.D. 1991. *et al*. World Bibliography of rice Stem Borers. *International Rice Research Institute, Los Banos, Philippines*, 1-426.

[45] Nagrajan S. Plant protection problems in rice-wheat rotation system: A perspective. Oryza. 1989; 26:329-33 (Original not seen. Abstr in CAB Abstracts AN: 19911151557).

[46] Paulian, F., & Popov, C. 1980. Sunn Pest or Cereal Bug (E. HAFLIGER editor). *Wheat, Ciba-Geigy, Basel*, 69-74.

[47] Brown, E. S., and M. Eralp. 1962. The distribution of the species of *Eurygaster* Lap. (Hemiptera, Scutelleridae) in Middle East countries. Annals and Magazine of Natural History 13(5):65-81.

[48] Javahery, M., C. W. Schaefer, and Lattin J. D. 2000. Shield bugs (Scutelleridae). Pages 475-486 *in* C. W. Schaefer and A. C. Panizzi (eds.). Heteroptera of Economic Importance. CRC Press, Boca Raton, FL.

[49] Popov, C., A. Barbulescu, and I. Vonica. 1996. Population dynamics and management of Sunn pest in Romania. Pages 47-59 *in* R. H. Miller and J. G. Morse (eds.). Sunn pests and their

control in the Near East. Food and Agriculture Organization of the United Nations, Rome.

[50] CABI, undated a. CABI Compendium: Status inferred from regional distribution. Wallingford, UK 2022.

[51] M. Jamil, A. Aziz, A. Ghaffar, H. Ali, S. Farooq, S. Ahmad and A. Zahid. 2014. Response of different wheat varieties/advanced lines to wheat aphids (Aphididae: Homoptera) and population trend of their natural predators. J. Pure Appl. Sci. 24-33:1-6.

[52] M. Razaq, W. Akhter, M. Faheem and F. Ahmad. 2005. Effect of sowing date of wheat on aphid (Schizaphis gramium RONDANI) population. Pak Entomol 27:79-82.

[53] Malipatil, M. 2008. Industry biosecurity plan for the grains industry threat contingency plan–Sunn pest–*Eurygaster integriceps*. Plant Health Australia, Canberra, Australia.

[54] Ali, W. K. and A. Q. S. Khidhir. 2016. Illustration of the morphologic characters of the sunn pest *Eurygaster integriceps* Puton, 1881 (Hemiptera: Scutelleridae) collected from Erbil Governorate-Kurdistan region-Iraq. Entomology, Ornithology & Herpetology 5(4):1-7.

[55] Ionescu, M. A. and C. Popov. 1976. Consideraţii asupra variabilitatii la *Eurygaster integriceps* (Heteroptera) [Considerations on variability in *Eurygaster integriceps* (Heteroptera)]. Studii şi cercetări de biologie 28(2): 89-94.

[56] Alexandrov N. 1947. *Eurygaster integriceps* Put. La varamine et ses parasites. *Entomologie et Phytopathologie Appliquees* 5: 29-41.

[57] Razaq, A. Hussain, M. Yaseen, M. Afzal and M. Mehmood. 2013. Yield and Yield Components of Wheat (Triticum Aestivum L.) affected by Aphid feeding and sowing time at Multan, Pakistan. *Pak.* J. Bot 45:2005-2011.

[58] Rezabeigi M, Esmaili M, Ganbalani GN, Radjabi G. 2000. Comparison of the greenhouse field methods in the host plant resistance experiments to the overwintered adults of sunn pest (*Eurygaster integriceps* Put.). In: *Proceedings of the 14th Plant Protection Congress of Iran*, 5-8 Sept. 2000, Vol. 1-Pests. page 3. Isfahan University of Technology.

[59] El Bouhssini, M., Street, K., Joubi, A., Ibrahim, Z., & Rihawi, F. 2009. Sources of wheat resistance to Sunn pest, Eurygaster integriceps Puton, in Syria. *Genetic Resources and Crop Evolution*, *56*(8), 1065.

[60] Karimzadeh, R., M. Hejazi, H. Helali, S. Iranipour and S. Mohammadi. 2011. Assessing the impact of site-specific spraying on control of *Eurygaster integriceps* (Hemiptera: Scutelleridae) damage and natural enemies. Precision Agriculture 12(4): 576-593.

[61] Kivan, M., and N. Kilic. 2006. Age-specific fecundity and life table of *Trissolcus semistriatus*, an egg parasitoid of the sunn pest *Eurygaster integriceps*. Entomological Science 9(1): 39-46.

[62] Logothetis, C. 1956. The Senn Pest, *Eurygaster integriceps*, in the Near East. FAO Plant Protection Bulletin 5(2): 21-25.

[63] Critchley, B. R. 1998. Literature review of sunn pest *Eurygaster integriceps* Put. (Hemiptera, Scutelleridae). Crop Protection 17(4): 271-287.

[64] EPPO, 2014. PQR Database. Paris, France: European and Mediterranean Plant Protection Organization.

[65] Donskoff M, 1996. Prospects for International Cooperation on Sunn Pest

Research and Control. FAO Plant Production and Protection Paper, 138:17-22

[66] Grigorov P, 1989. Effective damage caused by Eurygaster integriceps on wheat seeding quality. Rastenie Dni Nauki, 26(2):23-29

[67] Vinogradova N. M., 1969. Vredania ciripasca-Eurygaster integriceps. *In;* Trudi VIZR, 34, 98-133.

[68] El Sherif, S. I. 1972. On the biology of *Leucania loreyi* Dup. (Lepidoptera, Noctuidae). Zeitschrift für Angewandte Entomologie, *71*(1-4), 104-111.

[69] Hirai, K. and Santa H. 1983. Comparative physio-ecological studies on the armyworms, *Pseudaletia separata* Walker and *Leucania loreyi* Duponchel (Lepidoptera: Noctuidae). Bulletin of the Chugoku National Agricultural Experiment Station, E 83 No.21 pp.55-101 ref.75

[70] Burns A. N. and Mungomeryr, W. 1925. Investigations on sugarcane pests and diseases. Queensland Agric. J.I. 14 (4), 334-336.

[71] Sertkaya, E., & Bayram, A. 2005. Parasitoid community of the loreyi leaf worm Mythimna (Acantholeucania) loreyi: Novel host-parasitoid associations and their efficiency in the eastern mediterranean region of Turkey. Phytoparasitica, *33*(5), 441.

[72] Jiang, X. F., L. Zhang, Y. X. Cheng, and L. Z. Luo. 2014. Current status and trends in research on the oriental armyworm *Mythimna separata* (Walker) in China. Chinese J. Appl. Entomol. 51: 1444-1449.

[73] Guo, S. J., S. M. Li, L. P. Ma, and S. L. Li. 2001. Spatial distribution patterns and sampling techniques of larvae of *Leucania loreyi* Duponchel in corn fields. J. Henan Agric. Univ. 35: 245-248.

[74] Guo, S. J., S. M. Li, L. P. Ma, and X. N. Zhuo. 2003. Study on the biological characteristics and hazard lows of the *Leucania loreyi*. J. Henan Agric. Sci. 9: 37-39.

[75] Wu, R. Z. 1962. Preliminary study on *Leucania loreyi* Dup. Acta Entomol. Sin. 11: 164.

[76] Zhao, K. J. 1988. Preliminary study on occurrence regularity of *Leucania loreyi*. Chinese J. Appl. Entomol. 25: 140.

[77] Pike, K. S., Reed, G. L., Graf, G. T., & Allison, D. (1993). Compatibility of imidacloprid with fungicides as a seed-treatment control of Russian wheat aphid (Homoptera: Aphididae) and effect on germination, growth, and yield of wheat and barley. Journal of Economic Entomology, 86(2), 586-593.

[78] Archer, T. L. (1994, January). Economic injury levels and chemical control of the Russian wheat aphid. In Proceedings, Sixth Russian Wheat Aphid Workshop (pp. 23-25).

[79] Brenière, J. 1976. The principal insect pests of rice in West Africa and their control. *The principal insect pests of rice in West Africa and their control.*

[80] Karsholt, O., & Razowski, J. (Eds.). 1996. *The Lepidoptera of Europe: a distributional checklist*. Brill Academic Pub.

[81] Waterhouse, D. F. 1993. The major arthropod pests and weeds of agriculture in Southeast Asia: distribution, importance and origin (No. 435-2016-33732).

[82] Uhan, T. S. 1990. The biology of *Agrotis ipsilon* Hufn (Lepidoptera: Noctuidae) in the laboratory. *BuletinPenelitianHortikultura (Indonesia)*.

[83] Xiang, Y., Yang, M., Cui, W., Lou, Y., Tang, Y., & Li, Z. (2008). EAG

responses of the male black cutworm moth, *Agrotis ypsilon* (Rottemberg) (Lepidoptera: Noctuidae) to the female's sex pheromone. Acta Entomologica Sinica, 51(1), 91.

[84] Department of Primary Industries and Regional Development, Australia. 2020. Agriculture and Food, Cutworm: pests of crops and pastures.https://www.agric.wa.gov.au/pest-insects/cutworm-pests-crops-and-pastures.

[85] Shylesha, A., JALALI, S., GUPTA, A., VARSHNEY, R., VENKATESAN, T., SHETTY, P., OJHA, R., GANIGER, P.C., NAVIK, O. & SUBAHARAN, K. J. J. O. B. C. 2018. Studies on new invasive pest Spodopterafrugiperda (JE Smith) (Lepidoptera: Noctuidae) and its natural enemies. 32, 145-151.

[86] Kalleshwaraswamy, C. M., Asokan, R., Swamy, H. M., Maruthi, M. S., Pavithra, H. B., Hegbe, K., ... & Goergen, G. E. (2018). First report of the fall armyworm, Spodoptera frugiperda (JE Smith)(Lepidoptera: Noctuidae), an alien invasive pest on maize in India.

[87] (FAO) FOOD AND AGRICULTURE ORGANIZATION OF THE UNITED NATIONS. 2019c. First detection of fall armyworm in China. FAO. www.ippc.int/fr/news/first-detection-of-fall-armyworm-in-china/.

[88] Xu, Q., HATT, S., LOPES, T., ZHANG, Y., BODSON, B., CHEN, J. & FRANCIS, F. J. J. O. P. S. 2018. A push–pullstrategy to control aphids combines intercropping with semiochemical releases. 91, 93-103.

[89] Mchughen, A. & SMYTH, S. J. P. B. J. 2008. US regulatory system for genetically modified [geneticallymodified organism (GMO), rDNA or transgenic] crop cultivars. 6, 2-12.

[90] Say, T. 1831. Description of new species of heteropterous Hemiptera of North America. Description of new species of heteropterous Hemiptera of North America.

[91] Leonard, D. E. 1966. Biosystematics of the" Leucopterus complex" of the genus Blissus (Heteroptera: Lygaeidae). Connecticut Agricultural Experiment Station.

[92] Capinera, J. L. 2008. Encyclopedia of entomology. Springer Science & Business Media.

[93] Brandenburg, R. and J. Baker. 2020. Major Insect Pests of Turf in the US. Handbook of Integrated Pest Management for Turf and Ornamentals

[94] Spike, B., G. Wilde, T. Mize, R. Wright, and S. Danielson. 1994. Bibliography of the chinch bug, Blissus leucopterus leucopterus (Say) (Heteroptera: Lygaeidae) since 1888. Journal of the Kansas Entomological Society:116-125.

[95] Chakrabarti, S. 2018. Aphids. Pages 871-908 Pests and Their Management. Springer.

[96] Heie, O. E. 1967. Studies on Fossil Aphids.(Homoptera: Aphidoidea). Stiftsbogtr.

[97] Börner, C. 1952. Europae centralis Aphides: Die Blattläuse Mitteleuropas. Namen, Synonyme, Wirtspflanzen, Generationszyklen. Thüring. Botan. Ges.

[98] Heie, O. E. 1993. The Aphidoidea (Hemiptera) of Fennoscandia and Denmark. V, Family Aphididae, part 2 of tribe Macrosiphini of subfamily Aphidinae. Brill.

[99] Beirne, B. P. 1972. Pest insects of annual crop plants in Canada. IV. Hemiptera-Homoptera. V. Orthoptera. VI. Other groups. *Entomol Soc Can Mem*.

[100] Wilde, G. E., Whitworth, R. J., Claassen, M., Shufran, R. A. J. J. O. A. & Entomology, U. 2001. Seed treatment for control of wheat insects and its effect on yield. 18, 1-11

[101] Williams, I. S. and A. F. Dixon. 2007. Life cycles and polymorphism. Aphids as crop pests. Wallingford: CAB International:69-85.

[102] Sathe, T. V., A. Gophane, and N. Shendage. 2015. Colour attractivity and occurrence of some cell sap sucking pests on crop plants. Biolife**3**:540-546.

[103] Berlandier, F., D. Severtson, and P. Mangano. 2010. Aphid management in canola crops. Aphid management in canola crops.

[104] Schwertner, C. F., and J. Grazia. 2007. O genero Chinavia Orian (Hemiptera, Pentatomidae, Pentatominae) no Brasil, com chave pictorica para os adultos. Revista Brasileira de Entomologia 51: 416-435.

[105] Vivan, L. M., and A. R. Panizzi. 2002. Two new morphs of the southern green stink bug, *Nezara viridula* (L.) (Heteroptera: Pentatomidae), in Brazil. Neotropical Entomology 31: 475-476.

[106] Musolin, D. L. 2007. Insects in a warmer world: ecological, physiological and life-history responses of true bugs (Heteroptera) to climate change. Global Change Biology 13: 1565-1585.

[107] Panizzi, A. R., and T. Lucini. 2016. What happened to *Nezara viridula* (L.) in the Americas? Possible reasons to explain populations decline. Neotropical Entomology 45: 619-628.

[108] Tougou, D., D. L. Musolin, and K. Fujisaki. 2009. Some like it hot! Rapid climate change promotes shifts in distribution ranges of *Nezara viridula* and *N. antennata* in Japan. Entomologia Experimentalis et Applicata 30: 249-258.

[109] Musolin, D. L., and H. Numata. 2003. Photoperiodic and temperature control of diapause induction and colour change in the southern green stink bug *Nezara viridula*. Physiological Entomology 28: 65-74.

[110] McPherson, J. E., and R. M. McPherson. 2000. Stink bugs of economic importance in America north of Mexico. CRC Press, Boca Raton, FL. 253 pp.

[111] Prado, S. S., D. Rubinoff, and R. P. P. Almeida. 2006. Vertical transmission of a pentatomid caeca-associated symbiont. Annals of the Entomological Society of America 99: 577-585.

[112] Rojas, M. G., and J. A. Morales-Ramos. 2014. Juvenile coloration as a predictor of health in *Nezara viridula* (Heteroptera: Pentatomidae) rearing. Journal of Entomological Science 49: 166-175.

[113] Panizzi, A. R. 2000. Suboptimal nutrition and feeding behavior of hemipterans on less preferred plant food sources. Anais da Sociedade Entomologica do Brasil 29: 1-12.

[114] Chen, X. X., Feng, J. N., & Tong, X. L. (2011). Thrips (Insecta: Thysanoptera) of China. Check List, *7*, 720.

[115] Ananthakrishnan, T. N., & Thangavelu, K. (1976). The cereal thrips *Haplothrips ganglbaueri* Schmutz with particular reference to the trends of infestation onOryza sativa and the weedEchinochloa crusgalli. In *Proceedings of the Indian Academy of Sciences-Section B* (Vol. 83, No. 5, pp. 196-201). Springer India.

[116] Mound, L.A. & R. Marullo. 1996. The thrips of Central and South America: an introduction. Mems Entomol. 6: 1-487.

[117] Zhang, W.Q., X.L. Tong, X.N. Luo and W.X. Zhuo 1999. Thysanoptera; p. 347-395. *In* B.K. Huang (ed.). *Fauna of*

Insects Fujian province of China. Vol. I. Fuzhou: Science Technology of Fujian

[118] Loan, C., & Holdaway, F. G. (1955). Biology of the red clover thrips, Haplothrips niger (Osborn) (Thysanoptera: Phloeothripidae). The Canadian Entomologist, *87*(5), 210-219.

[119] Aheer, G.M., M. Ulfat, K. Jawad and A. Ali. 1993. Effect of sowing dates on aphids and grain yield in wheat. J. Agric. Res.

[120] Tabasum, S., I.R. Noorka, M. Afzal and A. Ali. 2012. Screening best adopted wheat lines against aphid (Schizaphis graminum Rondani) population. Pak. Entomol 34:51-53.

[121] Shahzad, M., H. Ghani, M. Ayyub, Q. Ali, H. Ahmad, A. Ali and M. Qasim. 2019. PERFORMANCE OF SOME WHEAT CULTIVARS AGAINST APHID AND ITS DAMAGE ON YIELD AND PHOTOSYNTHESIS. J. Glob. Innov. Agric. Soc. Sci. 105-109.

[122] Aslam, M. 2003. Population of Coccinella septempunctata L. in wheat planted on different dates. Pakistan Entomol. 25:45-48.

[123] Khan, M. and S.K. Khalil. 1990. Biological control of aphid with an entomopathogenic fungus. *Pakistan* J. Agric. Res. 11:174-177.

[124] Hayat, K., M.A. Chuhan, A. Rasul and I. Arshad. 2018. Intercropping of wheat and oilseed crops reduces wheat aphid, sitobion avenae (fabricius) (hemiptera: aphididae) incidences: a field study. J. Agric. Res. 56.

[125] Ahmed, N.E., H.O. Kanan, S. Inanaga, Y.Q. Ma and Y. Sugimoto. 2001. Impact of pesticide seed treatments on aphid control and yield of wheat in the Sudan. Crop Prot. 20:929-934.

[126] Ali, S., F. Akbar, A. Sultan and M. Saleem. 2018. An ecofriendly approach to control wheat aphid (schizaphis graminum (rondani) by using bio rational insecticides as seed treatment and foliar applications. 40:77-84.

[127] Suhail, A., J. Iqbal, M. Arshad, D. Gogi, M. Arif and T. Shafait. 2013. Comparative efficacy of insecticides as seed treatment against wheat aphid and its Coccinellid predator. 35:17-22.

[128] Irshad, M. 2001. Aphids and their biological control in Pakistan. *Pakistan* J. Biol. Sci. 4:537-541.

[129] Khan, S.A. and F. Ullah. 2005. Studies on the aphids distribution pattern and their natural enemies in wheat and maize crop. PhD thesis 61-62.

[130] Stary, P., K. Naumann-Etienne and G. Remaudiére. 1998. A review and tritrophic associations of aphid parasitoids (Hymenoptera, Braconidae, Aphidiinae) of Pakistan. *Parasit.*

[131] Khan, S.A., F. Ullah, N. Hussain, Y. Hayat and S. Sattar. 2007. Natural enemies of cereal aphids in North West Frontier Province (NWFP) of Pakistan. Sarhad J. Agric. 23:435.

[132] Zeb, Q., S. Rondon, H. Badshah and A. Khan. 2020. Influence of Cultivar on Aphids (Hemiptera: Aphididae) and Associated Natural Enemies in Pakistani Wheat Ecosystems. Pak. J. Zool. 52.

[133] Khan, H.A. and A. Suhail. 2001. Feeding Efficacy, Circadian Rhythms and Oviposition of the Lady Bird Beetle (Coccinellidae: Coleoptera) under Controlled Conditions. Int. J. Agric. Biol 3:384-386.

[134] Iqbal, J., M. Ashfaq and A. Ali. 2008. Management of aphids by augmentation of coccinellids and Chrysoperla carnea under field conditions on wheat. *Pakistan* J. Agric. Sci. 45:57-59.

[135] Faheem, M., S. Saeed, A. Sajjad, M. Razaq and F. Ahmad. 2019. Biological parameters of two syrphid fly species Ischiodon scutellaris (Fabricius) and Episyrphus balteatus (DeGeer) and their predatory potential on wheat aphid Schizaphis graminum (Rondani) at different temperatures. Egypt. J. Biol. Pest Control 29:2.

[136] Uddin, A., S. Ahmed, A. Ali, A. Khoso, M. Khan, F. Asghar and K. Asghar. 2019. Functional Response of Green Lacewing, Chrysoperla carnea (Neuroptera: Chrysopidae) Larvae on Different Insects Pests. 3:49-56.

[137] Farooq, Z., S. Fareed, H. Karar, M. Rubab and S.F.H. Shah. 2016. Efficacy of some botanical extracts against wheat aphids sitobion avenae (homoptera: aphididae) and their impact on predators population. J. Agric. Res 54:697-706.

[138] Aziz, A., M. Ahmad, M. Nasir and M. Naeem. 2013. Efficacy of Different Neem (Azadirachta indica) Products in Comparison with Imidacloprid against English Grain Aphid (Sitobion avenae) on Wheat. Int. J. Agric. Biol. 15:279-284.

[139] Abid, B. 2015. Toxicity of selected plant extracts against wheat aphid and its predators. Appl. Sci. Bus. Econ. 2:33-39.

[140] Iqbal, M.F., M.H. Kahloon, M.R. Nawaz and M.I. Javaid. 2011. Effectiveness of some botanical extracts on wheat aphids. J. Anim. Plant Sci. 21:114-115.

[141] Ullah, A., F. Muhammad and S. Shah. 2018. Influence of different sowing dates and wheat (Triticum aestivum) genotypes on aphid infestation and its predation in agro-climate of Lasbela region. J. Entomol. Zool. Stud. 6:2449-2453.

[142] Ajmal, M., J. Iqbal, M. Qayyum, M. Asad Saleem, M. Tayyab and M. Sajjad. 2018. Preferential influence of wheat varieties (Triticum aestivum L.) on population build-up of aphid (Homoptera: Aphididae) and its natural enemies. *J. Entomol.* Zool. Stud. 6.

Chapter 4

Xanthobacter autotrophicus an Endophytic Beneficial Bacterium for Wheat and Other Plants: A Short Review

Juan Manuel Sánchez-Yañez

Abstract

The endophytic genus plant growth promoting bacteria (EPGPB) known as *Xanthobacter autrotrophicus* is one of the most interesting option to apply on the production of wheat (*Triticum aestivum*), and other domestic crops lettuce (*Lactuca sativa*), tomato (*Solanum lycopersicum*) rice (*Oriza sativa*) maize (*Zea mays*): under all types of agriculture systems: open field, protecting one or either organic sustainable type. The aims of this review is to analyze the qualities of *X. autotrophicus* as useful EPGPB for sustainable production of wheat and other crops regarding its capacity as able to fix molecular nitrogen (N_2) as well as by transforming plant metabolic compounds in phytohormons, including phosphatase enzyme for solubilizing phosphate to solve different soil problems related with its fertility also some phytopathological like to stop of growing weed as *Arabidopsis thaliana* which are competing with health growth of domestic plants. Beside the potencial of *X. autotrophicus* for bioremediation of environmental polluted by chemicals.

Keywords: soil proprieties, cereals, vegetables and green nitrogen fertilizer, endophytic plant growth promoting bacteria, health

1. Introduction

The well know endophytic plant growth promoting bacterium *Xanthobacter autotrophicus*, was described as *Corynebacterium autotrophicus* due its specific genetic qualities to grow under chemolithoautotrophically and for being able to fix molecular nitrogen (N_2) as nitrogen source [1]: *X. autotrophicus* are rods, according to growth condition show pleomorphism depends on the species and the carbon and nitrogen source on which they are grown. *X. autotrophicus* is a Gram-type negative rods with high concentrations of polyphosphate granule belongs phylogenetically to the family Hyphomicrobiaceae in the class Alphaproteobacterial, grow heterotrophically under aerobic or microaerophilic conditions with acids, alcohols, and selectively with some carbohydrates as energy and carbon source like: fructose, galactose, mannose and sucrose [2–7].

The endophyte plant growth promoting bacteria: *X. autotrophicus* can fix dinitrogen under heterotrophic and thioautotrophic conditions is able to grow with

IntechOpen

Biochemical characteristics	1.	2.	3.
Cell morphology: rods	–	–	–
Morphology as rod on free carbon and nitrogen media	+	+	+
Slime production	+	+	+
Zeaxanthine dirhamnoside (yellow)	+	+	+
Zeaxanthine (orange, pinkish)	–	–	–
Motility under autotrophic growth conditions	–[d]	–[d]	–[d]
Vitamins required for growth	+	+	+
Sensitivity to chloramphenicol	–	–	–
Under autotrophic growth at 35°C	+	+	+
Utilization of hexoses	+	+	+
Growth on nutrient broth	+	+	+
Growth on glutamine as carbon source	+	+	+
Growth on citrate	+	+	+
Degradation of aromatic compounds	+	+	+
Degradation of cyclohexene (and derivatives)	+	+	+
Utilization of methanol	+	+	+
Utilization of hydrocarbons	+	+	+

Symbols and abbreviations: +, positive; (+), positive except for some unusual strains; –, negative; (–), negative except for some unusual strains; +/– not determined; TCA = tricarboxylic acid.[a]1 = X. autotrophicus in the reference; 2 = X. autotrophicus; 3 = X. autotrophicus repetitions.
[s]Lime production in glucose.
[c]Pale yellow indicating low concentration [6, 22–26].

Table 1.
Main biochemical characteristics among some species of the genus Xanthobacter.[a]

H_2 plus O_2 or H_2 + $Na_2S_2O_3$ as energy source and with CO_2 as only inorganic carbon source [8–10] at reduced O_2 tension [11, 12] in the absence of organic or inorganic nitrogen (N) compounds as aminoamides, peptides, proteins, nucleotides, well known sources as NH_4^+ (ammonia) or NO_3^- (nitrate) at the soil [13] and culture artificial media [14]. On the basis of their numbers, *X. autotrophicus* should be regarded as an associative symbiosis diazotroph due although entering roots of wheat (*Triticum aestivum*), bean (*Phaseolus vulgaris*), root beet (*Beta vulgaris*), rice (*Oriza sativa*) [15] tomato (*Solanum lycopersicum*), lettuce (*Lactuca sativa*) [16, 17] does not form nodules the way do symbiotic N_2-fixing bacteria in legume as *Bradyrhizobium japonicum* does. The special position of *X. autotrophicus* among the chemolithoautotrophic and other the N_2-fixing aerobic bacteria [1, 11], *X. autotrophicus* is able to grow with $H_2/CO_2/O_2$ or to have high hydrogenase activity [10, 18], beside reaction of nitrogenase as the other well-known genera: *Azotobacter*, *Derxia*, *Bradyrhizobium* and *Rhizobium* [19, 20]. Originally, one key taxonomic property for discriminating *X. autotrophicus* from other genera yellow pigmented zeaxanthin dirhamnoside bacteria, including diazotrophs [2]. *Xanthobacter* strains can be isolated easily if certain conditions are applied: no other or very limiting sources of nitrogen other than N_2 or $H_2/CO_2/O_2/N_2$ [9, 11] as gas phase providing an electron donor, a carbon source, electron acceptors, in liquid media; yellow colonies are showed on nutrient agar plates [1, 6]. Because its metabolic diversity *Xanthobacter* species are widespread in natural habitats [21] as is showed in **Table 1**.

2. Phylogeny and taxonomy of *Xanthobacter* spp

The phylogenetic position of *Xanthobacte*r based on 16S rRNA sequence analysis published in Bergey's Manual of Systematic Bacteriology, showed that genus Xanthobacter is part of phylum Proteobacteria, class Alpha proteobacteria order Rhizobiales family Xanthobacteraceae. However, using phylogenetic trees constructed on the basis of 16S rRNA sequence comparisons, the type strains of *Aquabacter spiritensis* and *Azorhizobium caulinodans* are intermingled with the otherwise well-defined genus cluster *Xanthobacter, Aquabacter* and *Azorhizobium* (both single species genera are recognized as separate genera V and VI within the same family Hyphomicrobiaceae, some of the key properties described for the type species *X. autotrophicus* it is suggested to keep the separate genera names despite the 16S rRNA sequence similarity [22]. The 16S rRNA sequence is more than 98%

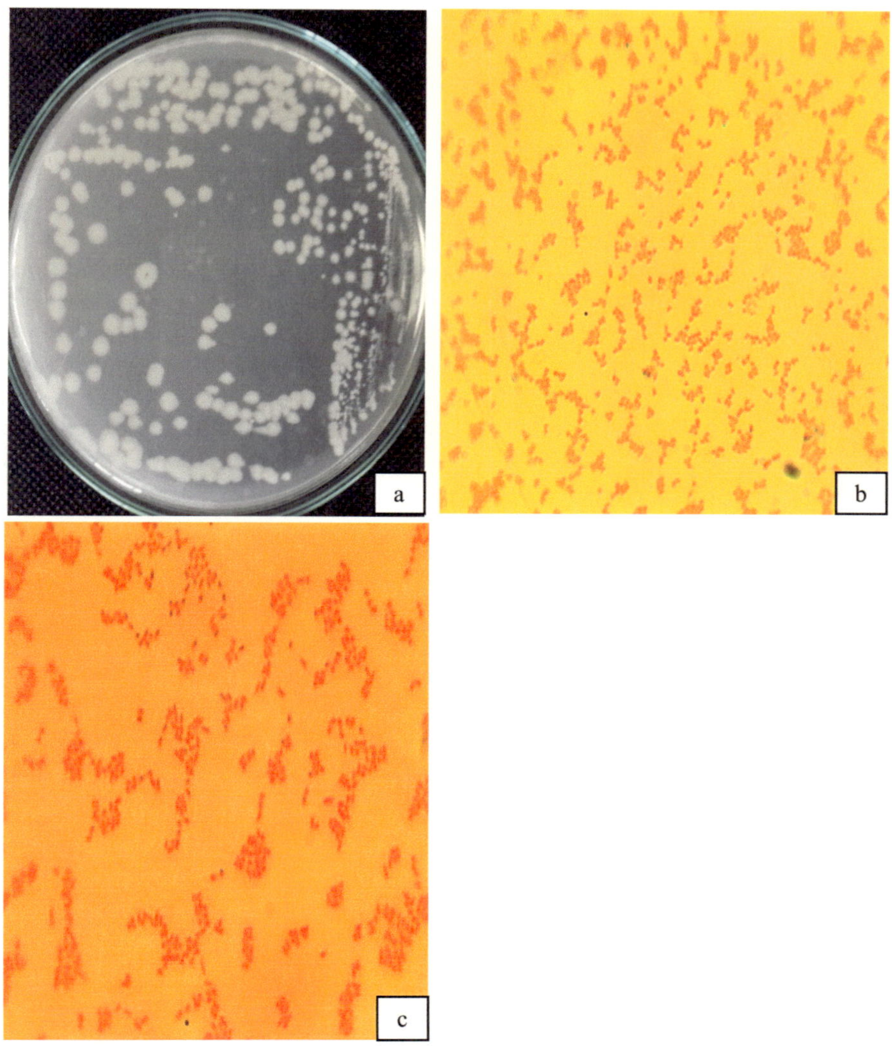

Figure 1.
Photographs of X. autotrophicus *(a) macroscopic morphology in a mineral medium without sucrose either ammonium nitrate (MMWSA) under autotrophic conditions after 30 h incubation at 35°C, (b) and (c) microscopic morphology of* X. autotrophicus *according at Gram negative in MMWSA under the same incubation condition (photos from Environmental Laboratory-UMSNH, Sánchez-Yañez et al., 2020).*

similar to those of *X. flavus* and *X. autotrophicus* strains as is showed in **Figure 1** [2, 6, 8, 23, 27]. The morphology and some of the physiological proprieties are different to separate species, supported by the low below 50% DNA-DNA hybridization data as well as tricarboxylic acid or TCA cycle intermediate; (1) synthesis of the water insoluble zeaxanthin dirhamnoside, showed by the yellow colonies; (2) normally to grow chemolithoautotrophically; and (3) able to fix dinitrogen under microaerophilic chemolithoautotrophic or heterotrophic conditions [1, 28]. Other characteristics are given in **Table 1**, *Xanthobacter* is free-living in soil and water as well as root-associated but never noduling, exhibits acetylene reduction as an indirect technique for nitrogen fixing capacity. Other features of *Xanthobacter* a are the h i g h G+C related with some flavobacteria and Cytophaga spp: (1) antibiotic pattern sensitivity [17] (2) positive reaction for catalase, oxidase and phosphatase acid and alkaline types; (3) negative reaction for methyl red, gas from carbohydrates, and the Voges-Proskauer test; and (4) containing ubiquinones Q10 and Q8 like is in *Beijerinckia*, and *Azotobacter*, are important for truly identification of these species, demonstration of the pigment zeaxanthin dirhamnoside and acquisition of the 16S rDNA sequence are important [2, 6, 21].

3. Taxonomy

The most identifications of environmental isolates are done by 16S rRNA sequence analysis, in a first common identification step, diagnostic taxonomic properties are: (1) yellow, "fried egg" shaped colonies with several amounts of slime production under cultivation media specific conditions; (2) rods, some species have strong polymorphic, branched, twisted cell morphology growing on nutrient agar with larger amounts of polyphosphate granula, can lead to the false impression of a Gram-positive staining reaction; however all *Xanthobacter* stain are truly Gram negative when is using a counterstain in polyphosphate-free cells of *X. autotrophicus* [2, 22, 24].

4. Isolation cultivation and axenic culture

Selective enrichment cultures. For isolation purposes, the use of free carbon and nitrogen agar medium as a selective medium is recommended for recovering *Xanthobacter* from; soil, upper layers of marine or freshwater sediments, lake water, steam and root of aal types of plants. Because of slime formation by *X. autotrophicus* of agar plates free carbon and nitrogen source. Frequently, other oligotrophic organisms grow as contaminants in the slimy colonies of *Xanthobacter* easy to separate in nutrient agar the following basal medium can be used for autotrophic as well as heterotrophic growth except when urea is used as nitrogen source. No vitamins or additions of yeast extract as growth factor are required for most *X. autotrophicus*, enrichment 100 mg of yeast extract per liter to the mineral medium can reduce an extended lag time for autotrophic growth under free carbon and nitrogen-fixing. In order to demostrate its capacity for fixing N2 is important not to add any inorganic or organic nitrogen source. During isolation, a vitamin solution any mixture containing biotin can enrichment to stimulate is growth, absence of ammonium, or amino acids, peptide, protein as any organic nitrogen source [29, 30]. For heterotrophic growth, common carbon sources are used: 0.5% sugars, 0.3% (v/v) alcohols or 0.4–0.8% organic acids. For growth under non-N_2-fixing conditions, 0.1% of ammonium chloride or sulfate is common. The

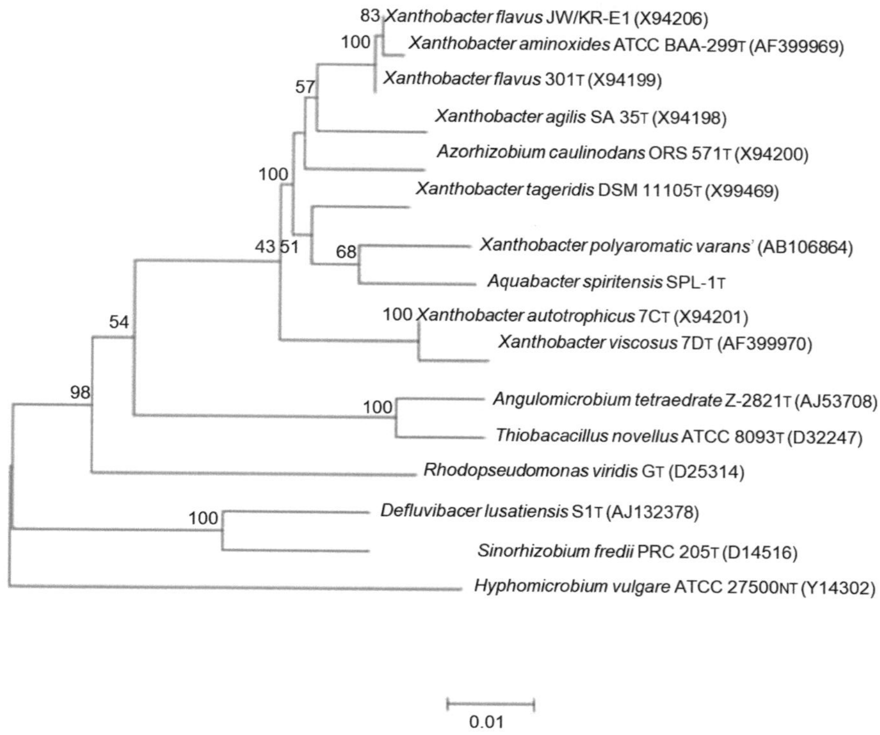

Figure 2.
Phylogeny and taxonomy of Xanthobacter *spp.*

exact composition of this medium is not critical, and good results have been obtained with free sucrose and nitrogen agar medium, for storge sterile soil is one the easy and best one to preserve viability for relative long period of time. *X. autotrophicus*, studied in more detail, most of the strains tested grow at pH 5.0–8.5 while pH recommend is about 6.8–7.2 and its temperature 30–37°C. The morphological features of *Xanthobacter* can be used initially for identification. Colony morphology depends on the type of carbon and nitrogen source and growth conditions. On most carbohydrates, the colonies of main species are large from 1 to 5 mm in diameter, smooth, convex, circular, filiform, opaque, and typically egg-yolk yellow color due to zeaxanthin dirhamnoside (see **Figure 2a**). The colonies become less yellow and less opaque as the amount of slime increases. The production of slime on nutrient agar plates frequently results in colonies resembling fried eggs [15]. Zeaxanthin dirhamnoside is water-insoluble, in contrast to the reddish/pinkish/brown pigment or to the yellow-green diffusing pigments with fluorescence of *Beijerinckia* and *Derxia* the other yellowish diazotroph isolated with well-known methods. The latter fact also makes it easy to distinguish *Xanthobacter* from *Derxia* colonies, which turn brown with age besides other morphological and biochemistry characteristics [5, 31]. *Xanthobacter* strains are sensitive to wide range of antibiotics, but the response depends on the method applied broth cultures or the use of Difco (Dispense-O-Disk minifilters). *X. autotrophicus* was sensitive to ampicillin, penicillin, chloramphenicol, erythromycin, novobiocin, and polymyxin B, but they were resistant to erythromycin and bacitracin. Few strains can grow on violet red-bile medium (Oxoid), deoxycholate medium (Oxoid), tellurate agar (Difco), and mineral medium supplemented with crystal violet red colonies [6, 32, 33].

4.1 Methods of storage

Xanthobacter cultures are grown on chemolithoautotrophic agar slants stored 1.5 years at 4°C after sealing the tubes tightly with parafilm. Also, liquid cultures grown under chemolithoautotrophic conditions mineral medium with 0.02% (w/v) yeast extract have been kept for more than 15 months at 4°C and, of course if glycerol is used as suspended solution at 40–60% (v/v) final, at –20°C and –75°C for more than 8 years. For long-term storage, cultures should be lyophilized on skim milk at 10% now sterile soil is an easy and safe technique [34, 35].

5. Autotrophy and nitrogen fixation capability

X. autotrophicus can use H_2 from thiosulfate as source of energy for CO_2 fixation, when grown heterotrophically in the presence of gas mixture, *Xanthobacter* species fix CO_2 mainly via the ribulose-biphosphate pathway but phosphoenol-pyruvate carboxylase activity also has been reported. Have shown that the fixation of CO_2 plays an important role in the degradation of aliphatic epoxides and ketones by novel carboxylases [5, 8, 10, 24]. *X. autotrophicus* fixes N_2 under heterotrophic growth conditions with sucrose as a carbon source; however, N_2 fixation was showed for several strains of *X. autotrophicus* with $15N_2$ incorporation into cell protein [12, 18]. The biochemical studies on the enzyme and its relationship to oxygen have been restricted to *X. autotrophicus*. The nitrogenase in these two strains is similar to that in other aerobic diazotrophs [2, 6, 36, 37]. There is strong variation among the strains in respect to the optimal O_2 concentration for growth under N_2-fixing conditions, for *X. autotrophicus*. The optimal partial pressures of O_2 for acetylene reduction are 5 and 2.5 kPa to 0.36 kPa. However, the alternative vanadium nitrogenase system could not yet be shown through substantial ethane synthesis or improving its growth when vanadium is added to molybdenum deprived medium [1, 14, 38].

6. Natural habitats

The known habitats of *Xanthobacter* are depending on its physiological properties, underline its catabolic versatility [39]. The sources for isolated strains include oil-contaminated soil and sludge from Japan [5], marine sediments, water and sediment samples from fresh- water lakes, soil of flooded rice fields, rhizosphere of wetland street ditches and wet meadow soil and garden soil from Europe, South Africa, North America, and Asia, sewage samples [3, 13, 40] and tree leaves [20, 41]. *Xanthobacter* is ubiquitous in microaerophilic environments with decaying organic material or matter [19] containing sufficient concentrations of H_2 and CO_2 and other metabolic compounds products of anaerobic microbial activity, such as organic acids and alcohols. *Xanthobacter* species are important in the microaerophilic interface between the anaerobic and aerobic habitats. Therefore, it is very likely that *Xanthobacter*, and possibly also other N_2-fixing Knallgas bacteria, can be found in habitats other not yet known [42, 43]. According to literature, no thermophilic, psychrophilic, or halophilic strains have been isolated [44]. Furthermore, it is not clear whether *Xanthobacter* contributes significantly as an associative bacterium to the nitrogen cycle in agriculture issues, even though in greenhouse experiments [17, 45, 46], *X. autotrophicus* strains isolated from several environmental samples have been shown to stimulate and growth yields of rice, tomato and lettuce at reduced dose of nitrogen or phosphate fertilizer [1, 17, 38, 47].

DOI: http://dx.doi.org/10.5772/intechopen.102066

6.1 Ecological interactions with other domestic plants

In Japan was reported a survey of N_2-fixing bacteria from roots of rice, with strains called group 2 were *X. autotrophicus* and other isolated *Xanthobacter*-like of group 5, which could be a new genus [22, 31, 48, 49]. Some of these isolations were identified as a *X. flavus* on the basis of morphological and physiological properties. Up to 25% of the nitrogen fixed by soil bacteria was incorporated into rice plants and other reported. In one soil like soils of Kasakh, Armenia *Xanthobacter* was up 40–70% were N_2-fixing population they may contribute to N balance in the soil of paddy rice. Also was demonstrated that strains close to *X. autotrophicus* could be found in the sediment of patty rice fields in Arkansas, United States, with more than 105 cells per g dry weight of roots in the rhizosphere of rice clearly as an endophyte [50]. A positive interaction among Xanthobacter and some domestic crops due to enhance biomass of plant as well as nitrogen content compared to those crops without Xanthobacter at limited dose of nitrogen fertilizer [17, 40]. Therefore, *Xanthobacter* can be classified as an associative diazotroph [19, 38, 44, 51, 52]. The possible role of *Xanthobacter* as a contributor of fixed N_2, a growth factor stimulant on bean [45] lettuce, tomato, rice, rootbeet, wheat, plants, and an associative N_2 fixer through either the phyllosphere or even stems nodules if in the future *Azorhizobium* is incorporated into the genus *Xanthobacter* needs to be investigate [20, 41, 53]. These studies should examine: (1) the role of the slime produced by *Xanthobacter* in its adherence to the rhizosphere and phyllosphere an involvement of slime in adherence processes was shown for various anaerobic bacteria; (2) the possible role of the polyglutamine polymer produced under high-nitrogen conditions directly after nitrogen fertilization [7, 38, 54] and (3) the role of plant growth stimulant formation by root and stem-associated *Xanthobacter* cells [13, 28, 55–57]. It has been reported than cultures of *X. autotrophicus* are producing indoleacetic acid when grown in medium with tryptophan [3]. Until now there are no reports about *Xanthobacter* isolated associated with any plant disease [18].

6.2 Biofertilizer application of endophytic plant growth promoting bacteria in modern sustainable agriculture

Biofertilizer is key action of organic farming and a main element for the economy in general modern agricultural production on a world scale [55, 56, 58, 59]. The biofertilizers play an important role in improving the fertility of the soil [60, 61]. In addition, their application in soil improves the structure of the soil minimizes the sole application of chemical fertilizer. Grain yield and harvest index also increase with use of biofertilizers. Inoculation with *Azotobacter* + *Rhizobium* + mycorrhizae gave the highest increase in straw and grain yield of wheat plants with rock phosphate as a P fertilizer. *Azolla* is inexpensive, economical, friendly, which provide benefit in terms of carbon and nitrogen enrichment of soil [62]. Some commercially available biofertilizers are also used for the crop. Raj [63] recorded that microorganism: *B. subtilis, Thiobacillus thioxidans*, and *Saccharomyces*) can be used as bio-fertilizers for solubilization of fixed micronutrients like Zn (zinc). As well for biological control, a modern approach of disease management a key role in sustainable agriculture [64–66]. Biofertilizers can be defined as carriers that contain living endophytic plant growth promoting bacteria (EPGPB) and/or microorganisms (EPGPM); when they are applied to seeds, plant surfaces, to soil, or to hydroponic agricultural system, they colonize the root system or interior of the plant, and to stimulate plant growth by increasing the demanding or availability of macro or micro minerals: nitrogen (N), phosphorus (P), potassium (K), cupper (Cu), iron (Fe), etc., to the host plant [67, 68]. According to Mishra et al. [69], biofertilizer could be mixture of active or latent microbial cell for several important mechanisms to improve plant growth and yield as the well-known: nitrogen fixing, phosphate solubilizing, or cellulolytic microorganisms for applications to soil, seed, roots, or composting involving any microbial process with the aims for enhancing plant growth, augment the availability of nutrients that can then be easily absorbed by the plants, as well as for biological control of plant pathogenic agents. Malusá and Vassilev [70] proposed that a biofertilizer is the formulated product containing one or more microorganisms that enhance the mineral availability for health growth and yield profitable performance of the plants by either replacing soil nutrients and/or by making nutrients more available to plants and/or by increasing plant availability to basic minerals [66, 71].

Biofertilizer products are usually based on the EPGPB or PGPM can be classified into three main types of microorganisms: arbuscular mycorrhizal fungi or AMF [72], plant growth promoting rhizobacteria or PGPR [73], and nitrogen fixing rhizobia and free nitrogen fixing bacteria for non-leguminous plant [74, 75] which are applied and approved as beneficial for domestic crops growth based in mineral nutritional, underline reported that PGPR are recommend worldwide as biofertilizers, contributing to maintain profitable yield without soil deterioration and preventing environmental pollution. Hence, with the potential contribution of the PGPR, to sustainable agriculture and forestry when pandemic condition of COVID 19 caused economic world depress [76, 77]. Sufficient densities of PGPR and/or EPGPB like *X. autotrophicus* in biofertilizer provide a beneficial role in creating a proper rhizosphere for plant growth and converting nutritionally important elements through biological process, for example increasing the availability of N, P, K, as well as inhibiting pathogens growth [67, 71].

The increasing availability of N, P, and K is enhancing soil fertility, to improve antagonistic capacity of PGPR or EPGPB to biocontrol of plant pathogens agents [58] as well as the survival time in all types of soil [78]. Previous studies show that a biofertilizer prepared by mixing all types of PGPR with composts or carriers could enhance growth- promoting effects and biocontrol of plants [79]. *Bacillus* spp [80] and *Pseudomonas* spp [81] are two PGPR that have been reported to effective biocontrol agents. Among these bacteria species, *Bacillus subtilis, B. amyloliquefacients*, and *B. cereus* are the most effective species for controlling plant diseases in domestic crops by several mechanisms [82]. Due endospores of the genus and species of *Bacillus* are tolerant to adverse environmental conditions allows PGPR, to survive and even to grow in a wide range of soils, thus facilitating the effective formulation of biofertilizer [83]. Based in this biochemicals qualities as well as the biorestauration of hyper fertilized or deteriorated soil [43, 74, 84, 85]. However, *X. autotrophicus* has many biological mechanisms to avoid environmental stress without any specific resistance structure a quality of this genus and specie [3, 21, 35] that has been useful to treat environments contaminated by chemical agents [86].

*6.2.1 Biofertilizer (*X. autotrophicus*) for bioremediation of environment polluted by chemical agents*

In that sense EPGPB (likes PGPM or/and PGPR can be classified as biofertilizers when they sustainable options to plant nourishment and enrichment source that would useful for bioremediation and/or phytoremediation (double actions plants and biofertilizer) for soil contaminated by chemical agents [87, 88]. There for *X. autotrophicus* is has been applied in bioaugmentation trials for cleaning up any environmental impacted by chemical agents [86] which due to powerful genetic capacity is able to degradate a wide range of chemical agents under several environmental conditions either soil and or water in that sense it been reported that *X. autotrophicus* is able to biodegradate of 1,2- dichloroethane (DCE) one of the largest chlorinated industrial chemical, most of it being used for synthesis of vinyl chloride and smaller amounts for ethylene diamine and other chemicals. It was also used as a solvent. Groundwater contamination is mainly due to leakages and improper waste disposal. *X. autotrophicus* can attack DCE by using some specific enzimes under oxic conditions was investigated in the 1980s [89, 90] required for prolonged groundwater bioremediation polluted by DCE. Such systems are operated under non-sterile conditions, and long-term survival of enzymes would require separate enzyme production and a process allowing for physical separation of the biocatalyst from groundwater. There may be attractive application opportunities if biotransformation of synthetic chemicals in waste streams leads to products that can be recycled, e.g., when a wastewater product can cleaning up. This issue that received attention during the development of strains of *X. autotrophicus* growing on 1,2,3-trichloropropane (TCP) and another xenobiotic compound that polluted wastewater [86, 90–92].

A bioformulation is not effective until it does not have an impact in field conditions, market existence and reliability and cost effectiveness [93]. Production of bioformulation is not only dependent on the detailed knowledge of microbial as well as plant physiology, but a number of technological challenges are also involved such as fermentation process, formulation type, population of microbe, and delivery systems [94]. Barea [59] has published that in order to get better bioformulation for any domestic crops is important to understand the interaction among EPGPB or PGPM. To reproduce those microorganisms is important the chemical composition of broth media as well as the main and best conditions for each microorganisms need to get enough amount of them for bioformulation applying in open agriculture [95]. Including legal and ecological permission for safe crops production. A key quality of any bioformulation has to be water soluble to make sure a positive effect on any domestic crop Himel et al. [96] and Bateman [97] underline for those bioformulation which are applying in in aerosol based on a droplet size that is sufficient to inoculate seeds and plants with excellent results. For bioformulations applied foliarly, it is important to consider all environmental factors: solar radiation, high temperatures, ultraviolet light, etc. that limit the survival of beneficial plant microorganisms [98]. In this sense, the type of bioformulations must be appropriate to the form and vehicle that transport the beneficial plant microorganisms according to the recommended application directly to the soil, to the seeds or plants so that the forecast of the result favors agricultural production or control of some disease or pest [99]. Therefore, it is important research for the innovation of bioformulation suitable for agricultural crops [58] that comply with the quality and legality standards to satisfy the world market demand for safe food without risk of environmental damage [100]. A fundamental aspect for the world market of biological inoculants has been the necessary implementation of microbiological quality controls with reliable protocols that are endorsed by laws in the world that protect those farmers who, when applying them, have the confidence that they will have positive results in production. agriculture, due in part to the unfortunate experience of bioformulations without microbiological or legal quality that have caused a rejection of some sectors involved in sustainable agricultural production, an aspect that has not yet been resolved in the world [68]. In an integral sense that the biotechnology of the formulation of inoculants requires solid research for the best selection of microorganisms that promote plant growth, as well as the protocols of legal and ethical microbiological quality in the generation of bioformulations that give confidence to be used in the world for a sustainable and harmless agricultural production in harmony with the environment [85].

7. Effect of *Xanthobacter autotrophicus* on the growth of *Triticum aestivum* and other domestic plants

Triticum aestivum is the main cereal consumed by the human population of the world, around 51% of human demand intake of calories and proteins. The annual production of this crop is ~630 million tons, being the major grown cereal worldwide with ~740 million ha harvested annually [101]. It is reported that the dynamics of colonization of endophytic genus plant growth promoting bacteria (EPGPB) like *X. autotrophicus* on the sphermosphere/rhizosphere in gramineae is reported, based in other genera and species different than *Xanthobacter* [49, 102, 103]. In that sense the response of domestic plants to *X. autotrophicus* is scarce [18] so research in progress indicates [16, 17, 45, 46] that it may be an excellent option for the sustainable production of domestic crops [67, 104–106] however it is believed that it may be similar to other genera and species of EPGB, of the known like *Azospirillum, Azotobacter, Bacillus* which are able to move from outside to into the root system [18, 107]. In the case of *T. aestivum* has a positive response to *X. autotrophicus* since can invade the interior of the root system where it transforms organic compounds derived from photosynthesis into phytohormons, to optimize the reduced dose of nitrogen fertilizer [46]. It has been showed that can invade the root of *T. aestivum* including other types of domestic crops [17, 45, 60]. This biochemical characteristic of *X. autotrophicus* was confirmed by its growth dependent on the nutritional richness of the rhizosphere of *T. aestivum*, attributable to certain organic acids, amino acids including other organic compounds from the photosynthesis in gramineae [42, 108]. Hence, the importance of the chemical composition of root, sphermosphere and rhizosphere, as inducers of colonization by *X. autotrophycus* in gramineae, is key for other EPGPB to be closely associated with its root, sphermosphere, rhizosphere system [60, 105, 109]; in part this also explains the nutritional requirement of *X. autotrophicus* for wheat as a distinctive characteristic of this species, which is not reported in *X. autotrophicus* this was verified when was inoculated in the soil without roots, this coupled with the competition and predation of the native soil microorganisms, antagonistic to the species of *Xanthobacter*, which prevented its persistence in that environment [20, 47]. In the literature it is reported that the positive response of *T. aestivum* to inoculation with *X. autotrophicus* and fed with NH_4NO_3 depends on fast they colonized exclusively the germination zone of the seed, as well as to invade inside the roots when they have developed [41, 109, 110]. This explains why, in the case of the test described, *X. autotrophicus* was detected during seed germination, in the period of root development, and even inside mature roots of wheat. This suggests that *X. autotrophicus* was not dependent on wheat's sphermosphere/rhizosphere [17, 108, 109], it is reported that slowly used its energy reserve to prolong its persistence in unsterilized soil, a physiological characteristic in *X. autotrophicus* [111, 112]. These results support that *T. aestivum* were attractive for *X. autotrophicus* used according to the type of root growth observed with *T. aestivum*, compared to the appearance of the root system in the coronary part and by the density of secondary roots detected uninoculated wheat [37, 46, 47, 113, 114].

Related to phosphorus a key mineral for plant nutrition as phosphates normally applied to soil as fertilizer it is reported that concentration in average soils is about 0.05% (w/w) of which only 0.1% is available to plants [115]. There is evidence that the phosphate fertilizer applied as phosphate has a limited impact on plant nutrition, especially because, due to the solubilization constant (Ksp), of this phosphate anion is generally little available for plant roots [116]. It is calculated in the soil the concentration of phosphorus as phosphates is equal to or less than 0.02ppm, which drastically limits plant growth [117, 118]. In nature, the strategy that plants use for the absorption of the forms of phosphates necessary for plant metabolism are

Figure 3.
Response of Triticum aestivum *to* Xanthobacter autotrophicus *at different levels of nitrogen and phosphate fertilizer at seedling stage 30 days after sowing. (a) Absolute control:* T. aestivum *not inoculated irrigated only with water; (b) relative control:* T. aestivum *not inoculated fed at 100% nitrogen and phosphate fertilizer; (c)* T. aestivum *with* X. autotrophicus *fed with 50% of nitrogen fertilizer and 100% phosphate fertilizer; (d)* T. aestivum *with* X. autotrophicus *fed with 100% nitrogen fertilizer and 50% phosphate fertilizer; (e)* T. aestivum *with* X. autotrophicus *fed with 50% nitrogen and phosphate fertilizer*

the solubilization actions of phosphates by genera and species of microorganisms that promote plant growth, such as mycorrhizae and bacteria that also mineralize organic compounds containing phosphates [119, 120]. In the last few years, the development of microbial inoculum containing phosphate-solubilizing microbes (PSM) gained attention of agriculturists [17].

Figure 3 shows the positive response of *T. aestivum* to *X. autotrophicus* fed at 50% of NH_4NO_3 and 100% phosphate fertilizer. **Figure 3c** shows that *T. aestivum* reached a greater number of leaves and a dense root system, as well as *T. aestivum* with *X. autotrophicus* fed with 100% nitrogen fertilizer and 50% phosphate fertilizer (**Figure 3d**) and *T. aestivum* with *X. autotrophicus* fed with 50% nitrogen and phosphate fertilizer (**Figure 3e**), compared to *T. aestivum* not inoculated irrigated with water (**Figure 3a**) and *T. aestivum* not inoculated fed with 100% nitrogen and phosphate fertilizer (**Figure 3b**). These facts indicates that *X. autotrophicus* transformed the organic compounds from photosynthesis of *T. aestivum* into root system to improve root absorption and optimize the reduced dose of nitrogen fertilizer without risk to plant health growth [17, 18, 38, 42, 43, 121, 122]. At the same time the synthesis of acid and mainly alkaline phosphatases improved the solubilization and absorption of soil phosphates and phosphate fertilizer apply [17, 123, 124] to enhance growth plant (data not showed). In **Figure 1** it was evident that *X. autotrophicus* is an excellent option for the sustainable production of *T. aestivum* since it is not only capable of optimizing nitrogen fertilizer to avoid soil deterioration and environmental contamination due to nitrogen hyperfertilization [36, 54, 121, 125, 126]. While *T. aestivum* inoculated with *X. autotrophicus* simultaneously absorbs the immobilized phosphate from the soil and optimizes the effective application from the inside of its roots by avoiding competition with the native microorganisms [13, 40, 53] with a high prognosis of achieving healthy growth and profitable yield [43, 122]. In that sense Khalid et al. [127] reported that seed inoculation with 30 bacterial strains isolated from rhizospheric soils of wheat plants cultivated at different sites significantly increased length and weight of wheat roots and shoots. Linear positive correlation between in vitro auxin production by these bacteria and increases in the measured growth parameters was observed. Abd El-Azeem et al. [128] reported a highly significant positive linear correlation between the in vitro auxin production by the tested PGPR strains and each of grain yield, straw and total yield (grain plus straw) as well as the number of tillers of wheat plants. Auxin or indole acetic acid (IAA) production is considered a way in which *X. autotrophicus* promotes plant growth by stimulating enzymological reactions [125, 129]. IAA influences plant processes, such as initiation of cell division and promotes vascular differentiation [130, 131]. Besides its hormonal functions, IAA is involved in the stimulation of ethylene synthesis, which is produced, by plants and microorganisms [47]. Ethylene plays several active roles in plants including germination of root and shoot and the response of plants to stress [43]. There is an evidence that *X. autotrophicus* that solubilize phosphate in soil and promote its uptake by plants are referred as phosphate solubilizing bacteria (PSB) or phosphobacteria and are included within EPGPB [132]. Plant growth promoting rhizobacteria increase the efficiency of fertilizers while reducing nitrogen loss. Their counts in the rhizosphere comprise a considerable share of the rhizospheric microorganisms and vary depending on the soil location and type as well as the cultivated plants [133]. Inoculating the soil or seeds with PSB individually or in combination with other microorganisms, especially the nitrogen-fixing bacteria increased the availability of P, Fe, Mn, Zn and Cu for plants and consequently increased crop yield [114, 134, 135].

Figure 4 shows the positive response of *Z. mays* to *X. autotrophicus* fed with 50% nitrogen fertilizer and 100% phosphate fertilizer (**Figure 4c**), had the highest number of leaves, plant height and the highest root density, as well as *Z. mays* with *X. autotrophicus* fed with 100% nitrogen fertilizer and 50% phosphate fertilizer (**Figure 4d**) and *Z. mays* with *X. autotrophicus* fed with 50% nitrogen and phosphate fertilizer (**Figure 4e**), compared to *Z. mays* not inoculated irrigated only with water (**Figure 4a**) and *Z. mays* not inoculated fed with 100% nitrogen and phosphate fertilizer (**Figure 4b**). **Figure 2** shows the effect of *X. autotrophicus* on the healthy growth of *Z. mays* at different doses of nitrogen and phosphorous fertilizer, supporting that *X. autotrophicus* from the interior of the root system of *Z. mays* had the ability to convert compounds generated from photosynthesis in phytohormones for the optimization of the fertilizer reduced to 50%, simultaneously with an increase in the acid and alkaline phosphatase activity for the solubilization of the immobile phosphates of the soil and the optimization of the phosphate fertilizer also reduced 50% [17, 44, 45, 47, 106, 131] compared to the limited growth of *Z. mays* without inoculation with *X. autotrophicus* where the absence of this endophytic bacterium

Figure 4.
Response of Zea mays *to* Xanthobacter autotrophicus *at different levels of nitrogen and phosphate fertilizer at seedling stage 15 days after sowing. (a) absolute control:* Z. mays *not inoculated irrigated with water; (b) relative control:* Z. mays *not inoculated fed with 100% nitrogen and phosphate fertilizer; (c)* Z. mays *with* X. autotrophicus *fed with 50% nitrogen fertilizer and 100% phosphate fertilizer; (d)* Z. mays *with* X. autotrophicus *fed with 100% nitrogen fertilizer and 50% phosphate fertilizer; (e)* Z. mays *with* X. autotrophicus *fed with 50% nitrogen and phosphate fertilizer.*

Figure 5.
Positive response of Oryza sativa *to* Xanthobacter autotrophicus *at different levels of nitrogen and phosphate fertilizer at seedling stage 15 days after sowing. (a) Absolute control:* O. sativa *not inoculated irrigated with water; (b) relative control:* O. sativa *not inoculated fed with 100% nitrogen and phosphate fertilizer; (c)* O. sativa *with* X. autotrophicus *fed with 50% nitrogen fertilizer and 100% phosphate fertilizer; (d)* O. sativa *with* X. autotrophicus *fed with 100% nitrogen fertilizer and 50% phosphate fertilizer; (e)* O. sativa *with* X. autotrophicus *fed with 50% nitrogen and phosphate fertilizer.*

that promotes plant growth shows that *Z. mays* that none of these fertilizers is efficiently absorbed, causing loss of soil fertility and a possible environmental contamination [43].

In **Figure 5**, showed the response of *O. sativa* to *X. autotrophicus* by the root length and plant height of *O. sativa* at 50% nitrogen fertilizer as NH_4NO_3 and 100% phosphorous fertilizer as K_2HPO_4/KH_2PO_4 (**Figure 5c**), in comparation to *O. sativa* with the maximum dose of nitrogen and phosphorous fertilizer but without *X. autotrophicus* (**Figure 5b**), as well as *O. sativa* with *X. autotrophicus* fed with 50% nitrogen and phosphate fertilizer (**Figure 5e**). This support that *X. autotrophicus* h is able to transform organic compounds from photosynthesis into phytohormons like auxins to increase root soil exploration for optimizing uptake of nitrogen fertilizer reduced to 50% [18, 45, 46, 106]. There is evidence that to support that *X. autotrophicus* in wheat, as well as in, oats, corn, sorghum, and other types of plants the way do other genus and species of growth plant promoting bacteria. While inside the roots of *O. sativa*; *X. autotrophicus* synthesizes acid and alkaline phosphatases for the solubilization and absorption of insoluble phosphate from the soil, as well as optimizing the phosphate fertilizer applied to the soil at a reduced dose without affecting the healthy growth of *O. sativa* compared to the response of *O. sativa* without inoculating with *X. autotrophicus* fed with the recommended dose of nitrogen and phosphate fertilizer, which shows that without the help of *X. autotrophicus*, *O. sativa* has growth limitations, therefore it is advisable to apply it to the sowing of the seed [119, 120, 123, 124, 126]. While inside the roots of *O. sativa*; *X. autotrophicus* synthesizes acid and alkaline phosphatases for the solubilization and absorption of insoluble phosphate from the soil, as well as optimizing the phosphate fertilizer applied to the soil at a reduced dose without affecting the healthy growth of *O. sativa* compared to the response of *O. sativa* without inoculating with *X. autotrophicus* fed with the recommended dose of nitrogen and phosphate fertilizer, which shows that without the help of *X. autotrophicus*; *O. sativa* has growth limitations, therefore it is advisable to apply it to the sowing of the seed [46, 106, 136].

In **Figure 6**, *L. sativa* inoculated with *X. autotrophicus* fed with 25% nitrogen and phosphate fertilizer (**Figure 6e**), as well as *L. sativa* with *X. autotrophicus* fed with 25% nitrogen fertilizer and 100% phosphate fertilizer (**Figure 6d**), had the highest

Figure 6.
Response of Lactuca sativa *to* Xanthobacter autotrophicus *at different levels of nitrogen and phosphate fertilizer at flowering stage 120 days after sowing. (a) Absolute control:* L. sativa *not inoculated irrigated with water; (b) relative control:* L. sativa *not inoculated fed with 100% nitrogen and phosphate fertilizer; (c)* L. sativa *with* X. autotrophicus *fed with 100% nitrogen fertilizer and 25% phosphate fertilizer; (d)* L. sativa *with* X. autotrophicus *fed with 25% nitrogen fertilizer and 100% phosphate fertilizer; (e)* L. sativa *with* X. autotrophicus *fed with 25% nitrogen and phosphate fertilizer; (f)* L. sativa *with* X. autotrophicus *fed with 0% nitrogen and phosphate fertilizer.*

number of leaves, plant height and the highest root density, compared with *L. sativa* not inoculated fed with 100% nitrogen and phosphate fertilizer (**Figure 6b**). It is reported that *X. autotrophicus* stimulated the proliferation of root hairs in wheat, as has been observed in other plants [17, 41, 44, 137, 138] and this increased the area of exploration of the root to capture the nitrogen and phosphate fertilizer [42], as reported in other works on *X. autotrophycus* inoculation: in corn [46, 106], in wheat and in rice [43, 139].

Figure 6 shows the effect of *X. autotrophicus* on the growth of *L. sativa* at different doses of nitrogen and phosphate fertilizer, where it was evident that *X. autotrophicus* can optimize the reduced dose of both fertilizers, in relation to nitrogen fertilizer by means of a conversion of metabolites released during photosynthesis [10, 17, 138], that reach the root to maximize the absorption of NH_4NO_3 while *X. autotrophicus* from inside the roots generates acid and especially alkaline phosphatases to solubilize the immobile phosphate of the soil, as well as optimize phosphate applied during the growth of *L. sativa* [140], in this trial it was demonstrated that these were the main mechanisms of *X. autotrophicus* when both fertilizers were applied in variable doses or in similar concentration, but not when in the absence of both [18, 120, 123, 124].

The possible synthesis of phytohormons by *X. autotrophicus* was supported by the test shown in **Figure 7**, in which it is evidenced by inoculation of *S. lycopersicum* with *X. autotrophicus* fed with 25% nitrogen and phosphate fertilizer (**Figure 7e**), had the highest number of leaves, fruits, plant height and the highest root density, compared to *S. lycopersicum* not inoculated fed with 100% nitrogen and phosphate fertilizer (**Figure 7b**). The positive growth of *S. lycopersicum* was due to the fact that *X. autotrophicus* had a growth promoter effect, which was detected from the beginning of wheat germination from its seed, reported to be maintained in the early stages of wheat root development [20, 41, 50], as observed in this experiment and which was similar to what was observed in root system when *X. autotrophicus* it colonizes and influences the growth of roots of beans [45, 54]. In that sense **Figure 7** shows the effect of *X. autotrophicus* on *S. lycopersicum* at different doses of nitrogen (NH_4NO_3) and phosphate (KH_2PO_4/K_2HPO_4) fertilizer, the growth of *S. lycopersicum* shows that the ability of *X. autotrophicus* to invade the interior of the radical system to transform compounds derived from photosynthesis into phytohormons improves the absorption and optimization of the reduced doses of NH_4NO_3 [13–15, 126] as well

Figure 7.
Response of Solanum lycopersicum *to* Xanthobacter autotrophicus *at different levels of nitrogen and phosphate fertilizer at maturity stage 180 days after sowing. (a) Absolute control:* S. lycopersicum *not inoculated irrigated with water; (b) Relative control:* S. lycopersicum *not inoculated fed with 100% nitrogen and phosphate fertilizer; (c)* S. lycopersicum *with* X. autotrophicus *fed with 100% nitrogen fertilizer and 25% phosphate fertilizer; (d)* S. lycopersicum *with* X. autotrophicus *fed with 25% nitrogen fertilizer and 100% phosphate fertilizer; (e)* S. lycopersicum *with* X. autotrophicus *fed with 25% nitrogen and phosphate fertilizer; (f)* S. lycopersicum *with* X. autotrophicus *fed with 0% nitrogen and phosphate fertilizer.*

as of phosphorous fertilizer, and even when none of them were applied to the crop is reported that N demand was supplied by biological N_2 fixation due to *X. autotrophicus* [18, 36, 52] it was also detected that it synthesized acid and alkaline phosphatase to solubilize the immobile of the soil, so that *S. lycopersicum* had a healthy growth with early formation of fruits [16, 17, 120, 135, 140] compared to *S. lycopersicum* without inoculating fed with the recommended doses both fertilizer.

Table 2 shows the acid and alkaline phosphatase activity of *X. autotrophicus*, measured indirectly by the amount of p-nitrophenol generated when measured in the stem and roots of *S. lycopersicum* (as it is a genus and endophytic growth plant promoting species) with nitrogen and phosphorous fertilizers at 25%, of the recommended dose, there it is observed that the values of the higher and lower acid

S. lycopersicum* (tomato)		**p-nitrophenol released (μg/ mL)**
Saline solution (absolute control)	Acid	—
	Alkaline	—
Without inoculating stem	Acid	0.45f**
	Alkaline	0.13f
Without inoculating root	Acid	1.49e
	Alkaline	0.16f
X. autotrophicus on the stem	Acid	140.22c
	Alkaline	102.66d
X. autotrophicus on the root	Acid	222.48a
	Alkaline	170.52b
X. autotrophicus isolated from the stem	Acid	139.77c
	Alkaline	102.53d

**n = 3.*
***Values with different letter are stadistically distint according to ANOVA-Tukey ($P < 0.05$).*

Table 2.
Activity of acid and alkaline phosphatases of Solanum lycopersicum at flowering stage 120 days after sowing at 25% of nitrogen and phosphate fertilizer with and without inoculating with Xanthobacter autotrophicus.

phosphatase of the alkaline support that the healthy growth of the vegetable was due to the activity of the phosphatases synthesized by *X. autotrophicus* not only the interior of the stem and better in the root, also when this strain of *X. autotrophicus* recovered from the stem as well as from the root results suggest the importance of soil phosphorus availability in altering the interactions between leading to soil invasion by *S. lycopersicum* by *X. autotrophicus*. Overall, applying high amounts of available nutrients may reduce and increase the abundance plant-beneficial microbes and pathobiome in soil, respectively, which in return, could affect soil and plant health. This work greatly advances the mechanistic understanding why *X. autotrophicus* is a genus with high competitive capacity within the broad group of growth-promoting endophytes that synthesize acid and/or alkaline phosphatases in the absence of available phosphates and even when soluble phosphates fertilizer is applied to soil in agricultural production [141], that issue could be important for researchers working in the field of environmental microbiology, microbial ecology, plant-microbe interactions, soil health, and plant protection [16–18, 123, 142] in comparison with the activity of both phosphatases of *S. lycopersicum* without inoculation with *X. autotrophicus*, where the poor activity of both phosphatases explains that the growth of this vegetable was not as vigorous as observed in *S. lycopersicum* inoculated with *X. autotrophicus* [120, 136]. Similar results of a high acid and alkaline phosphatase activity of *X. autotrophcius* inside the roots: *Beta vulgaris, Hordeum vulgare, Pinus leiophylla, T. aestivum, Sorhgum bicolor. Z. mays*, grown in soil with insoluble phosphate problems [124, 135, 137, 143] or precipitation of the phosphate fertilizer at a lower dose than recommended (data no shown).

Figure 8 shows that fruit of *S. lycopersicum* with *X. autotrophicus* fed with 100% nitrogen fertilizer and 25% phosphate fertilizer had the largest size and red coloration (**Figure 8**) while *S. lycopersicum* not inoculated irrigated fed with 100% nitrogen and phosphate fertilizer had a smaller size, in addition to a green coloration which means that vegetative life cycle was shorter than the fruit from not inoculate *S. lycopersicum* (**Figure 8a**). These results demonstrate the importance of *X. autotrophicus* for healthy growing plants, with a reduced dose of nitrogen and phosphorous fertilizer [54, 120, 144, 145]. **Figure 8** shows the effect of *X. autotrophicus* on the fruit of *S. lycopersicum* at a recommended dose of nitrogen fertilizer such as NH_4NO_3 with 25% of the phosphate fertilizer, in that sense *X. autotrophicus* is able to solubilize phosphate in soil and promote its uptake by plants are referred as phosphate solubilizing bacteria (PSB) or phosphobacteria and are included within PGPR [143]. Their counts in the rhizosphere comprise a considerable share

Figure 8.
Fruit of Solanum lycopersicum *with* Xanthobacter autotrophicus *at different levels of nitrogen and phosphate fertilizer at maturity stage 180 days after sowing. (a) Relative control:* S. lycopersicum *not inoculated fed with 100% nitrogen and phosphate fertilizer; (b)* S. lycopersicum *with* X. autotrophicus *fed with 100% nitrogen fertilizer and 25% phosphate fertilizer.*

of the rhizospheric microorganisms and vary depending on the soil location and type as well as the cultivated plants [133, 142, 146]. The results support that *X. autotrophicus* transformed organic compounds derived from photosynthesis in the inside the roots of *S. lycopersicum* in phytohormons for an efficient absorption of NH_4NO_3 while to optimize the phosphate fertilizer, *X. autotrophicus* by means of acid phosphatases, mainly alkaline phosphates solubilized the soil phosphates [123, 124, 142, 143, 145] and quickly absorbed the one applied consequently the fruit of the *S. lycopersicum* reached a larger size and ripened earlier in comparison with the size of the *S. lycopersicum* without inoculation fed with the recommended dose of both fertilizers [14, 16, 17, 38, 43, 50, 144].

Figure 9 shows that *A. thaliana* with *X. autotrophicus* fed with 100% NH_4NO_3 (**Figure 9d**), as well as *A. thaliana* with *X. autotrophicus* fed with 50% NH_4NO_3, (**Figure 9h**) and *A. thaliana* with *X. autotrophicus* irrigated with only water (**Figure 9l**) had root growth inhibition, its suggested due over synthesis of phytohormons not depending of NH_4NO_3 concentration [20, 79, 99, 131, 147] compared to *A. thaliana* with *B. vietnamiensis* 2 fed with 100% NH_4NO_3 (**Figure 9c**), as well as *A. thaliana* not inoculated fed with 50% NH_4NO_3 (**Figure 9e**) and *A. thaliana* not inoculated irrigated with water (**Figure 9i**).

Figure 10 shows that *A. thaliana* with *X. autotrophicus* fed with 100% NH_4NO_3 (**Figure 10d**), as well as *A. thaliana* with *X. autotrophicus* fed with 50% NH_4NO_3, (**Figure 8h**) and *A. thaliana* with *X. autotrophicus* irrigated with water (**Figure 10l**) had root growth inhibition, compared to *A. thaliana* with *B. vietnamiensis* 1 fed with 100% NH_4NO_3 (**Figure 10b**), as well as *A. thaliana* with *B. vietnamiensis* 1 fed with 50% NH_4NO_3 (**Figure 10f**) and *A. thaliana* with *B. vietnamiensis* 1 irrigated with water (**Figure 10j**). **Figures 9** and **10** show the response of the seed and stem primordia and root of *A. thalina* inoculated with *B. vietnamiensis* compared to *X. autotrophicus* at doses 100, 50 and 0% of the nitrogen fertilizer as NH_4NO_3 where it was evident that while a positive effect of *B. vietnamiensis* strains on *A. thaliana* was dependent on

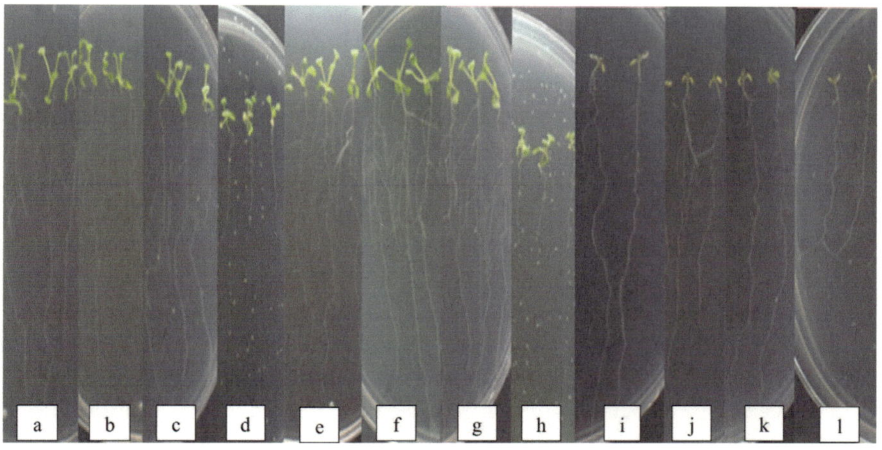

Figure 9.
Response of Arabidopsis thaliana *to* Burkholderia vietnamiensis *and* Xanthobacter autotrophicus *on the germination of seed and first step of growth at seedlings stage at different dose of NH_4NO_3 under artificial culture media. (a)* A. thaliana *not inoculated fed with 100% NH_4NO_3; (b)* A. thaliana *with* B. vietnamiensis *1 fed with 100% NH_4NO_3; (c)* A. thaliana *with* B. vietnamiensis *2 fed with 100% NH_4NO_3; (d)* A. thaliana *with* X. autotrophicus *fed with 100% NH_4NO_3; (e)* A. thaliana *not inoculated fed with 50% NH_4NO_3; (f)* A. thaliana *with* B. vietnamiensis *1 fed with 50% NH_4NO_3; (g)* A. thaliana *with* B. vietnamiensis *2 fed with 50% NH_4NO_3; (h)* A. thaliana *with* X. autotrophicus *fed with 50% NH_4NO_3; (i)* A. thaliana *not inoculated irrigated with water; (j)* A. thaliana *with* B. vietnamiensis *1 irrigated with only water; (k)* A. thaliana *with* B. vietnamiensis *2 irrigated with water; (l)* A. thaliana *with* X. autotrophicus *irrigated with only water.*

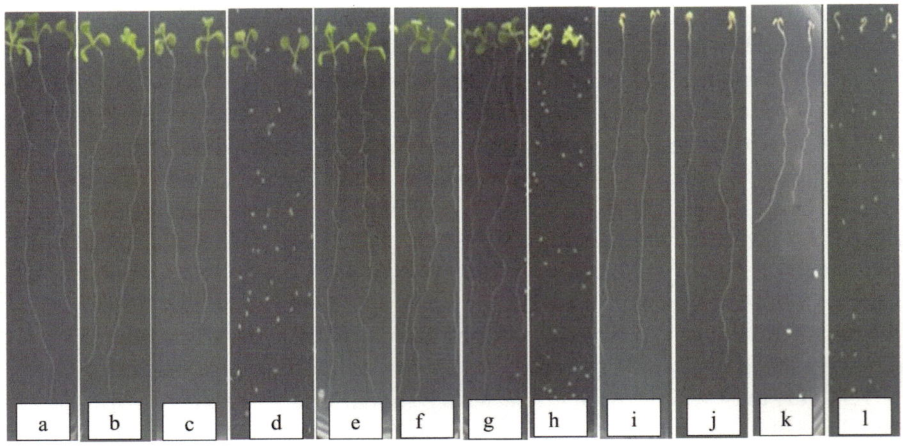

Figure 10.
Effect of Burkholderia vietnamiensis *and* Xanthobacter autotrophicus *on the germination of seed and first step of growth of* Arabidopsis thaliana *seeds directly sown in inoculated in artificial culture media at different dose of nitrogen fertilizer as NH4NO3. (a)* A. thaliana *not inoculated fed with 100% NH4NO3; (b)* A. thaliana *with* B. vietnamiensis 1 *fed with 100% NH4NO3; (c)* A. thaliana *with* B. vietnamiensis 2 *fed with 100% NH4NO3; (d)* A. thaliana *with* X. autotrophicus *fed with 100% NH4NO3; (e)* A. thaliana *not inoculated fed with 50% NH4NO3; (f)* A. thaliana *with* B. vietnamiensis 1 *fed with 50% NH4NO3; (g)* A. thaliana *with* B. vietnamiensis 2 *fed with 50% NH4NO3; (h)* A. thaliana *with* X. autotrophicus *fed with 50% NH4NO3; (i)* A. thaliana *not inoculated irrigated with water; (j)* A. thaliana *with* B. vietnamiensis 1 *irrigated with water; (k)* A. thaliana *with* B. vietnamiensis 2 *irrigated with water; (l)* A. thaliana *with* X. autotrophicus *irrigated with water.*

the concentration of NH_4NO_3, [41, 110, 129, 148] *X. autotrophicus* inhibited seed germination and practically stem and root primordium, both effects were positive by *B. vietnamiensis* well-known plant beneficial bacteria for a domestic vegetal [149]. In opposite way *X. autotrophicus* can distinguish between a domestic plant and a weed planted in agricultural soil by stimulating the growth of the former and inhibiting the germination and growth of the latter [150]. A genetic capacity that few genera and species such as *X. autotrophicus* of growth-promoting endophytic bacteria possess and can be used to improve the growth of domestic plants and prevent the germination of weeds underline when they are dependent on the synthesis of phytohormons from compounds releasing of the seed and roots of *A. thaliana* [46, 47, 59, 79, 131, 147].

8. Conclusions

The plant growth promoting endophytic bacteria well known as *X. autotrophicus* is an exceptional genus and species of procaryote due to the ability it has to simultaneously fix CO_2 and N_2, it can exist in water, soil and in association with a wide variety of plant species, specifically because by invading the root tissue it influences positively in the absorbing both nitrogen and phosphorous regulated forms of fertilization by the synthesis of acid and alkaline phosphatases in soil with phosphate availability problems or where the application of the phosphate fertilizer precipitates. Whereas by converting seed exudates and derivates and photosynthesis inside the root's plants in phytohormons have an interesting potential as a biofertilizer. As well as for the biological control of weeds that compete with the cultivation of domestic plants, it can contribute to sustainable agricultural production that reduces the effects of contamination by unregulated fertilization and application of chemical herbicides. Besides that *X. autotrophicus* is has been reported as useful biological tool for bioremediation of water and soil polluted by chemical agents. Easy to growth in simple culture media low cost to reproduce to industrial level, a friendly genus and bacterial species for humans, animals, plants and the environment.

Acknowledgements

The support of project 2.7 (2021-2022) of the Coordination of Scientific Research of the UMSNH, Morelia, Michoacan, Mexico is appreciated. As well as the project: "Field Test of a Living Biofertilizer for Crop Growth in Mexico" from Harvard University, Cambridge, Ma, USA with funding from the Rockefeller Foundation (2019–2022). To Jeaneth Caicedo Rengifo for her help working in this project. To Juan Luis Ignacio de la Cruz, MSc Blanca Celeste Saucedo Martinez and J Alberto Castro-Villaseñor for their technical support.

Conflict of interest

The authors declare no conflict of interest.

Author details

Juan Manuel Sánchez-Yañez
Environmental Microbiology Laboratory, Research Institute in Biology and Chemistry, Universidad Michoacana de San Nicolas de Hidalgo, Morelia, Mich, Mexico

*Address all correspondence to: syanez@umich.mx

References

[1] Hoffman BM, Lukoyanov D, Yang ZY, Dean DR, Seefeldt LC. Mechanism of nitrogen fixation by nitrogenase: The next stage. Chemical Reviews. 2014;**114**:4041-4062. DOI: 10.1021/cr400641x

[2] Willems A, Busse J, Goor M, Pot B, Falsen E, Jantzen E, et al. Hydrogenophaga, a new genus of hydrogen-oxidizing bacteria that includes Hydrogenophaga flava comb. nov. (formerly *Pseudomonas flava*), Hydrogenophaga palleronii (formerly *Pseudomonas palleronii*), Hydrogenophaga pseudoflava (formerly *Pseudomonas pseudoflava* and "*Pseudomonas carboxydoflava*"), and Hydrogenophaga taeniospiralis (formerly *Pseudomonas taeniospiralis*). International Journal of Systematic and Evolutionary Microbiology. 1989;**296** 1694-1697. DOI: 10.1099/00207712-39-3-319

[3] Doronina NV, Trotsenko YA. Reclassification of "*Blastobacter viscosus*" 7d and Blastobacter aminooxidans14a as *Xanthobacter viscosus* sp. nov. and *Xanthobacter aminoxidans* sp. nov. International Journal of Systematic and Evolutionary Microbiology. 2003;**53**: 179-182. DOI: 10.1099/ijs0.02231-0

[4] Euzéby JP. List of Bacterial Names with Standing in Nomenclature [Online]. 2004. Available from: http://www.bacterio.cict.fr/search.html [list updated frequently]

[5] Hirano SI, Kitauchi F, Haruki M, Imanaka T, Morikawa M, Kanaya S. Isolation and characterization of *Xanthobacter polyaromaticivorans* sp nov 127W that degrades polycyclic and heterocyclic aromatic compounds under extremely low oxygen conditions. Bioscience, Biotechnology, and Biochemistry. 2004;**68**:557-564

[6] Wiegel J. The Genus *Xanthobacter*. Bergey's Manual of Systematic Bacteriology. In: The Proteobacteria. New York, NY, USA: Springer-Verlag; 2004. DOI: 10.1007/0-387-30745-1_16

[7] Kramer C, Gleixner G. Variable use of plant- and soil-derived carbon by microorganisms in agricultural soils. Soil Biology and Biochemistry. 2006;**38**: 3267-3278. DOI: 10.1016/j.soilbio.2006.04.006

[8] Meijer WG, Arnberg AC, Enequist HG, Terpstra P, Lidstrom M, Dijkhuizen L. Identification and organization of carbon dioxide fixation genes in *Xanthobacter flavus* H4-14. Molecular and General Genetics. 1991;**225**:320-330

[9] Sluis MK, Ensign SA. Purification and characterization of acetone carboxylase from *Xanthobacter* strain Py2. PNAS. 1997;**94**:8456-8461. DOI: 10.1073/pnas.94.16.8456

[10] Liu C, Colón BC, Ziesack M, Silver PA, Nocera DG. Water splitting-biosynthetic system with CO_2 reduction efficiencies exceeding photosynthesis. Science. 2016;**352**:1210-1213. DOI: 10.11/science.aaf5039

[11] Ooyama J. Simultaneous fixation of CO_2 and N_2 in the presence of H_2 and O_2 by a bacterium. Report of the Fermentation Research Institute. 1971;**39**:41-44

[12] Nakamura Y, Yamanobe T, Ooyama J. Identification of nitrogen-fixing hydrogen bacterium strain N34 and its oxygen-resistant segregant strain, Y38. Agricultural and Biological Chemistry. 1985;**49**:1703-1709. DOI: 10.1080/00021369.1985.10866980

[13] Mäder P, Fliessbach A, Dubois D, Gunst L, Fried P, Niggli U. Soil fertility and biodiversity in organic farming. Science. 2002;**296**:1694-1697. DOI: 10.1126/science.1071148

[14] Murrell JC, Lidstrom ME. Nitrogen assimilation in *Xanthobacter* H4-14. Archives of Microbiology. 1983;**136**: 219-221

[15] Oyaizu-Masuchi Y, Komagata K. Isolation of free-living nitrogen-fixing bacteria from the rhizosphere of rice. The Journal of General and Applied Microbiology. 1988;**34**:127-164. DOI: 10.2323/jgam.34.127

[16] González Naranjo GO, Marquez-Benavides L, Sánchez-Yáñez JM. Respuesta de *Solanum lycopersicum* a *Xanthobacter autotrophicus* a diferentes niveles de fertilizantes nitrogenado y fosforado. In: 15 congreso Nacional de Ciencia Tecnología e Innovacion; 30 de octubre de 2020; Morelia, Michoacán, México (memoria in extenso) (in Spanish)

[17] Sanchez-Yañez JM, Castro VJA, Nocera D, Dogutan D, López Ortiz N. The México Innovation Fund (MIF). Progress Report Year 2. 2020. Field Test a Living Biofertilizer for Crop Growth in México (unpublished data)

[18] Liu C, Sakimoto KK, Colon BC, Silver PA, Nocera DG. Ambient nitrogen reduction cycle using a hybrid inorganic-biological system. PNAS. 2017;**114**(25):6450-6455. DOI: 10.1073/pnas.1706371114

[19] Chelius MK, Triplett EW. Prokaryotic nitrogen fixation: Model system for the analysis of a biological process. Wyndham, UK: Horizon Scientific Press; 2000. pp. 1-20

[20] Carvalho TLG, Balsemão-Pires E, Saraiva RM, Ferreira PCG, Hemerly AS. Nitrogen signalling in plant interactions with associative and endophytic diazotrophic bacteria. Journal of Experimental Botany. 2014;**65**: 5631-5642. DOI: 10.1093/jxb/eru319

[21] Wiegel J, Schlegel HG. Enrichment and isolation of nitrogen fixing hydrogen bacteria. Archives of Microbiology. 1976;**107**:139-142. DOI: 10.1007/BF00446833

[22] Rainey FA, Wiegel J. 16S rDNA sequence analysis confirms the close relationship between the genera *Xanthobacter, Azorhizobium,* and *Aquabacter* and reveals a lack of phylogenetic coherence among the species of the genus *Xanthobacter*. International Journal of Systematic Bacteriology. 1996;**46**:607-610. DOI: 10.1099/00207713-46-2-607

[23] Thompson JP, Skerman VBD. Azotobacteriaceae: The Taxonomy and Ecology of the Aerobic Nitrogen-fixing Bacteria. London, UK: Academic Press; 1979

[24] Urakami T, Araki H, Komagata K. Characteristics of newly isolated *Xanthobacter* strains and fatty acid compositions and quinone systems in yellow-pigmented hydrogen-oxidizing bacteria. International Journal of Systematic Bacteriology. 1995;**45**:863-867. DOI: 10.1099/00207713-45-4-863

[25] Reding HK, Croes GLM, Dijkhuizen L, Wiegel J. Emendation of *Xanthobacter flavus* as a motile species. International Journal of Systematic Bacteriology. 1992;**42**:309-311. DOI: 10.1099/00207713-42-2-309

[26] Reding HK, Wiegel J. Motility and chemotaxis of a *Xanthobacte*r rice isolate. Journal of General Microbiology. 1993;**139**:815-820. DOI: 10.1099/00221287-139-4-815

[27] Müller A, Schneider K, Erfkamp J, Wittneben V, Diemann E, Eaton AN. Vanadium-Akkumulationbeim Stickstoff-fixierenden Wasserstoffbakterium *Xanthobacter autotrophicus*. Naturwissenschaften. 1988;**75**:625-627

[28] Schneider K, Müller A, Krahn E, Hagen WR, Wassink H, Knüttel KH. The molybdenum nitrogenase from

wild-type *Xanthobacter autotrophicus* exhibits properties reminiscent of alternative nitrogenases. European Journal of Biochemistry. 1995;**230**:666-675. DOI: 10.1111/j.1432-1033.1995.0666h.x

[29] Shanmugam KT, Valentine RC. Microbial production of ammonium ion from nitrogen. Proceedings of the National Academy of Sciences of the United States of America. 1975;**72**: 136-139. DOI: 10.1073/pnas.72.1.136

[30] Ortiz-Márquez JCF, Do Nascimento M, Curatti L. Metabolic engineering of ammonium release for nitrogen-fixing multispecies microbial cell-factories. Metabolic Engineering. 2014;**23**:154-164. DOI: 10.1016/j.ymben.2014.03.002

[31] Reding HK. Ecological, Physiological, and Taxonomical Studies of Xanthobacter Strains Isolated from the Roots of Wetland Rice [PhD dissertation]. USA: University of Georgia; 1991

[32] Wiegel J. Distinction between the Gram reaction and the Gram type of bacteria. International Journal of Systematic Bacteriology. 1981;**31**:88. DOI: 10.1007/BF00446833

[33] Trower MK, Buckland RM, Higgins R, Griffin M. Isolation and characterization of a cyclohexane-metabolizing *Xanthobacter* sp. Applied and Environmental Microbiology. 1985;**49**:1282-1289. DOI: 10.1128/am.49.5.1289.1985

[34] Malik KA. A new freeze-drying method for the preservation of nitrogen-fixing and other fragile bacteria. Journal of Microbiological Methods. 1988;**8**:259-271. DOI: /10.1016/0167-7012(88)90008-5

[35] Malik KA. A simplified liquid-drying method for the preservation of microorganisms sensitive to freezing and freeze-drying. Journal of Microbiological Methods. 1990;**12**: 125-132. DOI: 10.1016/0167-7012(90)90022-X

[36] Dixon R, Kahn D. Genetic regulation of biological nitrogen fixation. Nature Reviews. Microbiology. 2004;**2**:621-662

[37] Pankievicz VC, do Amaral FP, Santos KFDN, Agtuca B, Schueller YMJ, Arisi ACM, et al. Robust biological nitrogen fixation in a model grass-bacterial association. The Plant Journal. 2015;**81**:907-919. DOI: 10.1111/tpj.12777

[38] Dashti N, Prithiviraj B, Hynes RK, Smith DL. Root and rhizosphere colonization of soybean (*Glycine max* L. Merr.) by plant-growth-promoting rhizobacteria at low root zone temperatures and under short-season conditions. Journal of Agronomy and Crop Science. 2000;**185**:15-22. DOI: 10.1046/j.1439-037X.2000.00394.x

[39] Geisen S, Mitchell EAD, Adl S, Bonkowski M, Dunthorn M, Ekelund F, et al. Soil protists: A fertile frontier in soil biology research. FEMS Microbiology Reviews. 2018;**42**:293-323. DOI: 10.1093/femsre/fuy006

[40] Samanta R, Sen SP. Further observations on the utility of N_2-fixing microorganisms in the phyllosphere of cereals. Journal of Agriculture Science (Cambridge). 1986;**107**:673-680. DOI: 10.1017/S0021859600069823

[41] Bent E, Tuzun S, Chanway CP, Enebak S. Alterations in plant growth and in root hormone levels of lodgepole pines inoculated with rhizobacteria. Canadian Journal of Microbiology. 2001;**47**:793-800. DOI: 10.1139/w01-080

[42] Christiansen-Weniger C, Groneman AF, Van Veen JA. Associative N_2 fixation and root exudation of organic acids from wheat cultivars of different aluminum tolerance. Plant and Soil. 1992;**139**:167-174

[43] Chaparro JM, Sheflin AM, Manter DK, Vivanco JM. Manipulating the soil microbiome to increase soil health and plant fertility. Biology and Fertility of Soils. 2012;**48**:489-499. DOI: 10.1007/s00374-012-0691-4

[44] Chanway CP, Shishido M, Nairn J, Jungwirth S, Markham J, Xiao G, et al. Endophytic colonization and field responses of hybrid spruce seedling after inoculation with plant growth-promoting rhizobacteria. Forest Ecology and Management. 2000;**133**:81-88. DOI: 10.1016/S0378-1127(99)00300-X

[45] Flores MJA, Ocampo J, Castro-Villaseñor JA, Sánchez-Yañez JM. Respuesta de *Phaseolus vulgaris* a *Xanthobacter autotrophicus* y *Rhizobium phaseoli*. In: 14 Congreso Estatal de Ciencia Tecnología e Innovacion del Estado de Michoacán. Morelia, Michoacán, 3 de octubre de 2019, México (memoria in extenso) (in Spanish)

[46] Sanchez-Yañez JM, Ocampo J, Nocera D, Dogutan D, López Ortiz N. The México Innovation Fund (MIF). Progress Report Year 1. 2019. Field Test a Living Biofertilizer for Crop Growth in México (unpublished data)

[47] Benizri E, Baudion E, Guckert A. Root colonization by inoculated plant growth-promoting rhizobacteria. Biocontrol Science and Technology. 2001;**11**:557-574. DOI: 10.1080/09583150120076120

[48] Padden AN, Rainey FA, Kelly DP, Wood AP. *Xanthobacter tagetidis* sp. nov., an organism associated with Tagetes species and able to grow on substituted thiophenes. International Journal of Systematic Bacteriology. 1997;**47**:394-401. DOI: 10.1099/00207713-47-2-394

[49] Wieland G, Neumann R, Backhaus H. Variation of microbial communities in soil, rhizosphere, and rhizoplane in response to crop species, soil type, and crop development. Applied and Environmental Microbiology. 2001;**67**:5849-5854. DOI: 10.1128/AEM.6712.58.49-5854.2001

[50] Rudrappa T, Czymmek KJ, Paré PW, Bais HP. Root secreted malic acid recruits beneficial soil bacteria. Plant Physiology. 2008;**148**:1547-1556. DOI: 10.1104/pp.108.127613

[51] Brown GD, Rovira AD. The rhizosphere and its management to improve plant growth. Advances in Agronomy. 1999;**66**:1-102. DOI: 10.1016/S0065-2113(08)60425-3

[52] James EK. Nitrogen fixation in endophytic and associative symbiosis. Field Crops Research. 2000;**65**:197-209. DOI: 10.1016/S0378-4290(99)00087-8

[53] Samanta R, Dutta AK, Sen SP. The utilization of leaf wax by N_2-fixing microorganisms on the leaf surface. Journal of Agriculture Science (Cambridge). 1986;**107**:681-685. DOI: 10.1017/S0021859600069835

[54] Wubs ERJ, van der Putten WH, Bosch M, Bezemer TM. Soil inoculation steers restoration of terrestrial ecosystems. Nature Plants. 2016;**2**:1026. DOI: 10.1038/NPLANTS2016107

[55] Kent AD, Triplett EW. Microbial communities and their interactions in soil and rhizosphere ecosystems. Annual Review of Microbiology. 2002;**56**: 211-236. DOI: 10.1146/annurev.micro.56.012302.161120

[56] Boraste A. Biofertilizers: A novel tool for agriculture. International Journal of Microbiology Research. 2009;**1**:23-31

[57] Ljung K. Auxin metabolism and homeostasis during plant development. Development. 2013;**140**:943-1016. DOI: 10.1242/dev.086363

[58] Hynes RK, Boyetchko SM. Research initiatives in the art and science of

biopesticide formulations. Soil Biology and Biochemistry. 2006;**38**:845-849. DOI: 10.1016/j.soilbio.2005.07.003

[59] Barea JM. Future challenges and perspectives for applying microbial biotechnology in sustainable agriculture based on a better understanding of plant–microbiome interactions. Journal of Soil Science and Plant Nutrition. 2015;**15**:261-282. DOI: doi.org/10.4067/S0718-95162015005000021

[60] Kachroo D, Razdan R. Growth, nutrient uptake and yield of wheat (*Triticum aestivum*) as influenced by biofertilizers and nitrogen. Indian Journal of Agronomy. 2006;**51**(1):37-39

[61] Son TN, Thu VV, Duong VC, Hiraoka H. Effect of Organic and Biofertilizers on Soybean and Rice Cropping System. Tsukuba, Ibaraki, Japan: Japan International Research Center for Agricultural Sciences; 2007

[62] Kaushik BD, Prassana R. Status of biological nitrogen fixation by cianobacteria and Azolla. In: Dadarwal KR, Yadav KS, editors. Biological Nitrogen Fixation Research Status in India: 1889-1989. New Delhi: Society of Plant Physiologist and Biochemists; 1989. pp. 141-208

[63] Raj SA. Bio-fertilizers for micronutrients. Biofertilizer Newsletter. 2007:8-10

[64] Hoffmann-Hergarten S, Gulati MK, Sikora RA. Yield response and biological control of Meloidogyne incognita on lettuce and tomato with rhizobacteria. Zeitschrift fur Pflanzenkrankheiten und Pflanzenschutz. 1998;**105**(4):349-358

[65] Senthilkumar T, Rajendran G. Biocontrol agents for the management of desease complex involving root-knot nematode, *Meloidogyne incognita* and *Fusarium moniliforme* on grapevine (*Vitis vinifera*). Indian Journal of Nematology. 2004;**34**(1):49-51

[66] Bernal P, Eberl L, de Jonge R, Lepek VC, Malone JG. Understanding plant-microorganism interactions to envision a future of sustainable agriculture. Environmental Microbiology. 2021;**23**:1809-1811. DOI: 10.1111/1462-2920.15479

[67] Vessey KJ. Plant growth promoting rhizobacteria as biofertilizers. Plant and Soil. 2003;**255**:571-586. DOI: 101023/A: 1026037216893

[68] Arora NK. Plant Microbe Symbiosis: Applied Facets. New Delhi, India: Springer; 2015. p. 383. DOI: 10.1007/978-81-322-2068-8

[69] Mishra D, Rajvir S, Mishra U, Kumar SS. Role of bio-fertilizer in organic agriculture: A review. Research Journal of Recent Sciences. 2013;**2**:39-41

[70] Malusá E, Vassilev N. A contribution to set a legal framework for biofertilisers. Applied Microbiology and Biotechnology. 2014;**98**:6599-6607. DOI: 10.1007/s00253-014-5828-y

[71] Waddington SR. Organic matter management: From science to practice. Soil Fertility. 1998;**62**:24-25

[72] Jeffries P, Gianinazzi S, Perotto S, Turnau K, Barea JM. The contribution of arbuscular mycorrhizal fungi in sustainable maintenance of plant health and soil fertility. Biology and Fertility of Soils. 2003;**37**:1-16. DOI: 10.1007/s00374-002-0546-5

[73] Podile AR, Kishore GK. Plant growth-promoting rhizobacteria. In: Plant-Associated Bacteria. Dordrecht, The Netherlands: Springer; 2006. pp. 195-230. DOI: 10.1007/978-1-4020-4538-7_6

[74] Chary SP, Schmid M, Hartmann A. Diversity of 16S- rRNA and nifH genes derived from rhizosphere soil and roots of an endemic drought tolerant grass, *Lasiurus sindicus*. European Journal of

Soil Biology. 2009;**45**:114-122. DOI: 10.1016/j.ejsobi.2008.06.005

[75] Franche C, Lidstrom K, Elmerich C. Nitrogen-fixing bacteria associated with leguminous and non-leguminous plants. Plant and Soil. 2009;**321**:35-59. DOI: 10.1007/s11104-008-9833-8

[76] Khalid A, Arshad M, Shaharoona B, Mahmood T. Plant growth promoting rhizobacteria and sustainable agriculture. In: Microbial Strategies for Crop Improvement. Berlin/Heidelberg, Germany: Springer; 2009. pp. 133-160. DOI: 10.1007/978-3-642-01979-1_7

[77] Thapa S, Sotang N, Adhilkari J, Ghimire A, Limbu AR, Joshi A, et al. Impact of COVID-19 lockdown on Agriculture Education in Nepal: An online survey. Pedagogical Research. 2020;**5**:4. DOI: 10.29333/pr/8465

[78] Yang X, Chen L, Yong X, Shen Q. Formulation can affect rhizosphere colonization and biocontrol efficiency of *Trichoderma harzianum* SQR-T037 against *Fusarium* wilt of cucumbers. Biology and Fertility of Soils. 2011;**47**: 239-248. DOI: 10.1007/s00374-010-0527-z

[79] Chen LH, Tang XM, Raze W, Li JH, Liu YX, Qiu MH, et al. *Trichoderma harzianum* SQR-T037 rapidly degrades allelochemicals in rhizospheres continuously cropped cucumbers. Applied Microbiology and Biotechnology. 2011;**89**:1653-1663. DOI: 10.1007/s00253-010-2948-x

[80] Gong M, Wang JD, Zhang J, Yang H. Study of the antifungal ability of *Bacillus subtilis* strain PY-1 in vitro and identification of its antifungal substance (Iturin A). Acta Biochimica et Biophysica Sinica. 2006;**38**:233-240. DOI: 10.1111/j.1745-7270.2006.00157.x

[81] Leonardo D, Blanca LF, Landa B, Weller DM. Host crop effects rhizosphere colonization and competitiveness of 2,4-diacetylphloroglucinol-producing *Pseudomonas fluorescens*. Phytopathology. 2006;**96**:751-762. DOI: 10.1094/PHYTO-96-0751

[82] Francis I, Holsters M, Vereecke D. The Gram-positive side of plant-microbe interaction. Environmental Microbiology. 2010;**12**:1-12. DOI: 10.1111/j.1462-2920.2009.01989.x

[83] Perez-Garcia A, Romero D, de Vicente A. Plant protection and growth simulation by microorganism: Biotechnological applications of *Bacillus* in agriculture. Current Opinion in Biotechnology. 2011;**22**:187-193. DOI: 10.1016/j.copbio.2010.12.003

[84] Requena N, Pérez-Solis E, Azcón-Aguilar C, Jeffries P, Barea JM. Management of indigenous plant-microbe symbioses aids restoration of desertified ecosystems. Applied and Environmental Microbiology. 2001;**67**:495-498. DOI: 10.1128/AEM.67.2.495-498.2001

[85] Mishra J, Arora NK. Bioformulations for plant growth promotion and combating phytopathogens: A sustainable approach. In: Bioformulations: For Sustainable Agriculture. India: Springer; 2018. DOI: 10.1007/978-81-322-2779-3_1

[86] Janssen DB, Stucki G. Perspectives of genetically engineered microbes for groundwater bioremediation. Environmental Science: Processes & Impacts. 2020;**22**:487. DOI: 10.1039/C9EM00601J

[87] Diab A. Phytoremediation of oil contaminated desert soil using the rhizosphere effects. Global Journal of Environmental Research. 2008;**2**(2):66-73

[88] Abioye OP, Agamuthu P, Aziz AA. Phytotreatment of soil contaminated with used lubricating oil using *Hibiscus cannabinus*. Biodegradation. 2012;**23**(2): 277-286. DOI: 10.1007/s10532-011-9506-9

[89] Stucki G, Krebser U, Leisinger T. Bacterial growth on 1,2-dichloroethane. Experientia. 1983;**39**:1271-1273

[90] Janssen DB, Scheper A, Dijkhuizen L, Witholt B. Degradation of halogenated aliphatic compounds by *Xanthobacter autotrophicus* GJ10. Applied and Environmental Microbiology. 1985;**49**:673-677. DOI: 10.11 28/aem.49.3.673-677.1985

[91] Tardif G, Greer CW, Labbé D, Lau PC. Involvement of a large plasmid in the degradation of 1,2-dichloroethane by *Xanthobacter autotrophicus*. Applied and Environmental Microbiology. 1991;**57**:1853-1857. DOI: 10.1128/aem.57.6.1853-1857.1991

[92] Stroo A, Leeson A, Ward CH. Bioaugmentation for Groundwater Remediation. New York, USA: Springer; 2013. DOI: 10.1007/978-1-4614-4115-1_4

[93] Brar SK, Verma M, Tyagi RD, Valero JR. Recent advances in downstream processing and formulations of *Bacillus thuringiensis* based biopesticides. Process Biochemistry. 2006;**41**:323-342. DOI: 10.1016/j.procbio.2005.07.015

[94] Malusá E, Sas-Paszt L, Ciesielska J. Technologies for beneficial microorganisms inocula used as biofertilizers. Scientific World Journal. 2012;**2012**:491206. DOI: 10.1100/2012/491206

[95] Montazeri M, Greaves MP. Effects of culture age, washing and storage conditions on desiccation tolerance of *Colletotrichum truncatum* Conidia. Biocontrol Science and Technology. 2002;**12**:95-105. DOI: 10.10.80/09583150120110680

[96] Himel CM, Loats H, Bailey GW. Pesticide sources to the soil and principles of spray physics. Pesticides in the Soil Environment: Processes, Impacts and Modeling. 1990;**2**:7-50

[97] Bateman R. Simple, standardized methods for recording droplet measurements and estimation of deposits from controlled droplet applications. Crop Protection. 1993;**12**: 201-206. DOI: doi.org/10.1016/0261-2194(93)90109-V

[98] Bailey P, Baker G, Caon G. Field efficacy and persistence of *Bacillus thuringiensis* var. kurstaki against Epiphyas postvittana (Walker) (Lepidoptera: CAPS PRA: Epiphyas postvittana 7 Tortricidae) in relation to larval behaviour on grapevine leaves. Aust. Journal of Entomology. 1996;**35**: 297-302. DOI: doi.org/10.1111/j.1440-6055.1996.tb01407.x

[99] Peng G, Wolf TM. Synergy between synthetic and microbial herbicides for weed control. Pest Technology. 2011;**5**:18-27

[100] Greaves J, Grant W. Under performing policy networks: The biopesticides network in the United Kingdom. British Politics. 2010;**5**:14-40

[101] Shiferaw B, Smale M, Braun HS, Duveiller E, Reynolds M, Muricho G. Crops that feed the world 10: Past success and future challenges to the role played by wheat in global food security. Food Security. 2013;**5**:219-317. DOI: 10.1007/s12571-013-0263-y

[102] Gray EJ, Smith DL. Intracellular and extracellular PGPR: Commonalities and distinctions in the plant-bacterium signaling processes. Soil Biology and Biochemistry. 2005;**37**:395-412. DOI: 10.1016/j.soilbio.2004.08.030

[103] Vejan P, Abdullah R, Khadiran T, Ismail S, Nasrulhaq Boyce A. Role of plant growth promoting rhizobacteria in agricultural sustainability—A review. Molecules. 2016;**21**(5):573. DOI: 10.3390/molecules21050573

[104] Nowak J. Review benefits of in vitro "bacterization" of plant tissue

cultures with microbial inoculants. In Vitro Cellular & Developmental Biology. Plant. 1998;**34**:122-130

[105] Lucy M, Reed E, Glick BR. Applications of free-living plant growth-promoting rhizobacteria. Antonie Van Leeuwenhoek. 2004;**86**:1-25

[106] Castro-Villaseñor JA, Velázquez Medina A, González Naranjo GO, Sánchez-Yáñez JM. *Xanthobacter autotrophicus* en *Zea mays* opción biológica para reducir fertilizante nitrogenado y fosforado. In: IV Coloquio de la Maestría en Ciencias en Ingeniería Ambiental. Universidad Michoacana de San Nicolás de Hidalgo; 4 de diciembre de 2020; Morelia, Michoacán, México (memoria in extenso). In Spanish

[107] Egamberdiyeva D, Hőflich G. Influence of growth-promoting bacteria on the growth of wheat in different soils and temperatures. Soil Biology and Biochemistry. 2003;**35**:973-978. DOI: 10.1016/S0038-0717(03)00158-5

[108] Amara MAT, Dahdoh MSA. Effect of inoculation with plant growth promoting rhizobacteria (PGPR) on yield and uptake of nutrients by wheat grown on sandy soil. Egyptian Journal of Soil Science. 1997;**37**:467-484

[109] Bashan Y, Levanony H. Interaction between *Azospirillum brasilense* and wheat root cells during early stages of root colonization. In: Klingmuller W, editor. *Azospirillum* IV. Berlin, Germany: Springer-Verlag; 1998. p. 166-173. DOI: 10.1007/978-3-642-73072-6_21

[110] Sarwar M, Arshad M, Martens DA, Frankenberger WT Jr. Tryptophan-dependent biosynthesis of auxins in soil. Plant and Soil. 1992;**147**:207-215

[111] Raymond J, Siefert SL, Staples CR. The natural history of nitrogen fixation. Molecular Biology and Evolution. 2004;**21**:541-559. DOI: 10.1093/molbev/msh047

[112] Ravindra S, Deep CS, Santosh S, Reeta G. Exploration of nifH gene through soil metagenomes of the western Indian Himalayas. 3 Biotech. 2016;**6**:25. DOI: 10.1007/s13205-015-0324-3

[113] Baldani VLD, Baldani JI, Döbereiner J. Inoculation of rice plants with the endophytic diazotrophs Herbaspirillum seropedicae and *Burkholderia* spp. Biology and Fertility of Soils. 2000;**30**:485-491. DOI: 10.1007/s003740050027

[114] Zaidi A, Ahmad E, Khan MS, Saif S, Rizvi A. Role of plant growth promoting rhizobacteria in sustainable production of vegetables: Current perspective. Scientia Horticulturae. 2015;**193**(1):231-239. DOI: 10.1016/j.scienta.2015.07.020

[115] Achal V, Savant VV, Reddy MS. Phosphate solubilization by a wild type strain and UV-induced mutants of *Aspergillus tubingensis*. Soil Biology and Biochemistry. 2007;**39**(2):695-699. DOI: 10.1016/j.soilbio.2006.09.003

[116] Holford ICR. Soil phosphorus: Its measurement, and its uptake by plants. Australian Journal of Soil Research. 1997;**35**:227-239. DOI: 10.1071/S96047

[117] Batjes NH. A world data set of derived soil properties by FAO–UNESCO soil unit for global modelling. Soil Use and Management. 1997;**13**:9-16. DOI: 10.1111/j.1475-2743.1997.tb00550.x

[118] Vassilev SV, Vassileva CG. A new approach for the classification of coal fly ashes based on their origin, composition, properties, and behavior. Fuel. 2006;**86**: 1490-1512. DOI: 10.1016/j.fuel.2006.11.020

[119] Fankem H, Nwaga A, Dueubel A, Dieng L, Merbach W, Etoa FX. Occurrence and functioning of phosphate solubilizing microorganisms from oil palm tree (*Elaeis guineensis*) rhizosphere

in Cameroon. African Journal of Biotechnology. 2006;**5**:2450-2460

[120] Khan MS, Zaidi A, Wani PA. Role of phosphate-solubilizing microorganisms in sustainable agriculture. A review. Agronomy for Sustainable Development. 2007;**27**: 29-43. DOI: 10.1051/agro:2006011

[121] Glick BR. The enhancement of plant growth by free-living bacteria. Canadian Journal of Microbiology. 1995;**41**:109-101. DOI: 10.1139/m95-015

[122] Barazani O, Friedman J. Is IAA the major root growth factor secreted from plant-growth-mediating bacteria? Journal of Chemical Ecology. 1999;**25**: 2397-2406. DOI: 10.1023/A:1020890311499

[123] Richardson AI. Prospects for using microorganisms to improve the acquisition of phosphorus y plants. Australian Journal of Plant Physiology. 2001;**28**:817-906. DOI: 10.1071/PP01093

[124] Richardson AE, Barea JM, McNeil AM, Prigent-Combart C. Acquisition of phosphorus and nitrogen in the rhizosphere and plant promoting by microorganisms. Plant and Soil. 2009;**321**:305-339. DOI: 10.1007/s11104-009-9895-2

[125] Khalid A, Arshad M, Zahir ZA, Khaliq A. Potential of plant growth promoting rhizobacteria for enhancing wheat yield. Journal of Animal and Plant Sciences. 1997;**7**:53-56

[126] Dos Reis FB Jr, Reis VM, Urquiaga S, Döbereiner J. Influence of nitrogen fertilization on the population of diazotrophic bacteria Herbaspirillum spp. and *Acetobacter diazotrophicus* in sugar cane (*Saccharum* spp.). Plant and Soil. 2000;**219**:153-159

[127] Khalid A, Arshad M, Zahir ZA. Screening plant growth-promoting rhizobacteria for improving growth and yield of wheat. Journal of Applied Microbiology. 2004;**96**:473-480. DOI: 10.1046/j.1365-2672.2003.02161.x

[128] Abd El-Azeem SAM, Mehana TA, Shabayek AA. Some plant growth promoting traits of rhizobacteria isolated from Suez Canal region, Egypt. African Crop Science Conference Proceedings. 2007;**8**:1517-1525

[129] Kravchenko LV, Borovkov AV, Pshikvil Z. The possibility of auxin biosynthesis in wheat rhizosphere by associated nitrogen-fixing bacteria. Mikrobiologiya. 1991;**60**:927-931

[130] Martens DA, Frankenberger WT Jr. Assimilation of exogenous 2-14C-indole acetic acid and 3-14C-tryptophan exposed to the roots of three wheat varieties. Plant and Soil. 1994;**166** 281-290

[131] Asghar H, Zahir Z, Arshad M, Khaliq A. Relationship between in vitro production of auxins by rhizobacteria and their growth-promoting activities in *Brassica juncea* L. Biology and Fertility of Soils. 2002;**35**(4):231-237. DOI: 10.1007/s00374-002-0462-8

[132] Paungfoo-Lonhiennea C, Reddingb M, Prattc C, Wang W. Plant growth promoting rhizobacteria increase the efficiency of fertilizers while reducing nitrogen loss. Journal of Environmental Management. 2019;**233**: 337-341. DOI: 10.1016/j.jenvman.2018.12.052

[133] Abd El-Azeem SAM. Influence of phosphate solubilizing bacteria in soil and rhizosphere on phosphorus availability [thesis]. Ismailia, Egypt: Faculty of Agriculture, Suez Canal University; 1998

[134] Ibrahim AN, Mahmoud AHH, El-Katkat MB. In vivo, study the effect of phosphate solubilizing bacteria and superphosphate on the productivity of broad bean and wheat plants. Menofiya

Journal of Agriculture Research. 1995;**20**:2361-2377. DOI: 10.1071/S96047

[135] Mehana TA, Farag FM. Influence of phosphate-dissolving microorganisms and elemental sulphur on phosphorus and micronutrient availability in a calcareous soil treated with rock phosphate. Mansoura University, Journal of Agricultural Sciences. 2000;**25**:2983-2993

[136] Berendsen RL, Pieterse CMJ, Bakker PAHM. The rhizosphere microbiome and plant health. Trends in Plant Science. 2012;**17**:478-486. DOI: 10.1016/j.tplants.2012.04.001

[137] Gyaneshwar P, Kumar GN, Parekh LJ, Poole PS. Role of soil microorganisms in improving P nutrition of plants. Plant and Soil. 2002;**245**:83-93

[138] González Naranjo GO, Márquez-Benavides L, Sánchez-Yáñez JM. Efecto positivo de *Xanthobacter autotrophicus* en *Lactuca sativa* a dosis reducida de fertilizante nitrogenado. In: IV Coloquio de la Maestría en Ciencias en Ingenieria Ambiental. Universidad Michoacana de San Nicolas de Hidalgo, 4 de diciembre 2020b, Morelia, Michoacán, México (memoria in extenso). In Spanish

[139] Reding HK, Hartel PG, Wiegel J. Effect of *Xanthobacter*, isolated and characterized from rice roots on growth of Wetland rice. Plant and Soil. 1991;**138**:221-229

[140] Arun KS. Bio-fertilizers for sustainable agriculture. Mechanism of P Solubilization. 6th ed. Jodhpur, India: Agribios publishers; 2007. pp. 196-197

[141] Pengfa L, Ming L, Guilong L, Kai L, Tianshun L, Meng W, et al. Phosphorus availability increases pathobiome abundance and invasion of rhizosphere microbial networks by Ralstonia. Environmental Microbiology. 2021;**23**(8):5992-6003. DOI: 10.1111/1462-2920.15696

[142] De Freitas JR, Banerjee MR, Germida JJ. Phosphate solubilizing rhizobacteria enhance the growth and yield but not phosphorus uptake of canola (*Brassica napus* L.). Biology of Fertility Soils. 1997;**24**:358-364. DOI: 10.1099/ijs.02231-0

[143] Gaind S, Gaur AC. Thermotolerant phosphate solubilizing microorganisms and their interaction with mungbean. Plant and Soil. 1991;**133**:141-149

[144] Simon HM, Smith KP, Dodsworth JA, Guenthner B, Handelsman J, Goodman RM. Influence of tomato genotipe on growth of inoculated and indigenous bacteria in the spermosphere. Applied and Environmental Microbiology. 2001;**67**:514-520. DOI: 10.1128/AEM.67.2.514-520.2001

[145] Afzal AFTAB, Ashraf M, Asad SA, Farooq M. Effect of phosphate solubilizing microorganisms on phosphorus uptake, yield and yield traits of wheat (*Triticum aestivum* L.) in rainfed area. International Journal of Agriculture and Biology. 2005;7(2):207-209

[146] El-Azeem A, Mehana T, Shabayek A. Effect of seed inoculation with plant growth-promoting rhizobacteria on the growth and yield of wheat (*Triticum aestivum* L.) cultivated in a sandy soil. Catrina: The International Journal of Environmental Sciences. 2008;**3**(2):69-74

[147] Guadarrama-Morales T, Mellado-Rojas ME, Carrillo-Flores E, Beltran-Peña E, Sanchez-Yañez JM. Effect of bacteria isolated from polluted soil by waste motor oil on root system architecture of *Arabidopsis thaliana*. In: XVIII National Congress of Biochemistry and Plant Molecular Biology XI Symposium, México, USA 1st ASPB. 2019

[148] Sarwar M, Kremer RJ. Enhanced suppression of plant growth through production of L-tryptophan-derived compounds by deleterious rhizobacteria. Plant and Soil. 1995;**172**:261-269

[149] De Sa OAAL, Ribeiro PRD, Rufini M, Cruvinel IAD, Casagrande DR, Moreira FMD. Microsymbionts of forage peanut under different soil and climate conditions belong to a specific group of Bradyrhizobium strains. Applied Soil Ecology. 2019;**143**:201-212. DOI: 10.1016/j.apsoil.2019.07.018

[150] Schlatter D, Kinkel L, Thomashow L, Weller D, Paulitz T. Disease suppressive soils: New insights from the soil microbiome. Phytopathology. 2017;**107**:1284-1297. DOI: 10.1094/PHYTO-03-17-0111-RVW

Chapter 5

Russian Wheat Aphid Distribution in Wheat Production Areas: Consequences of Management Practices

Astrid Jankielsohn

Abstract

Russian wheat aphid (RWA) is an international pest on wheat and occurs in most countries where large scale wheat cultivation is practiced. Consequently, considerable efforts have been made to manage RWA globally. The two management options used currently are chemical control and breeding for deployment of resistant wheat cultivars. There are however drawbacks to both of these management practices. Chemical control has a negative impact on the environment, especially other insect groups such as predators, pollinators and decomposers. With widespread and continuous use of the same active ingredients, there is the possibility that RWA can build up resistance against these specific active ingredients. The drawback with resistance breeding is that certain RWA populations can overcome the resistance in the wheat, resulting in new biotypes virulent to the resistant wheat cultivars.

Keywords: Russian wheat aphid, *Diuraphis noxia*, wheat, *Triticum aestivum*, biotypes, insecticide resistance

1. Introduction

Establishment success and rate of spread will determine the invasive ability of a specific organism [1]. The success of an invasive species will further be determined by both abiotic and biotic factors that will influence the adaptation and spread within the geographic range of establishment [2]. Liu *et al* [3] believe that Russian wheat aphid (RWA), *Diuraphis noxia* (Kurdjumov) possesses many of the features that define a 'good invader' and as a result became a global threat to wheat production. RWA has originally spread from central Asia [4] to other major wheat (*Triticum aestivum* L.) producing countries in the world. It is considered a primary pest of dryland winter wheat in North America [5] and South Africa [6]. RWA, like other exotic aphid species, is capable of surviving at low numbers for a relatively long period and can have sudden population outbreaks in new areas [7]. The most recent record of this aphid invading a new area was in 2016 in Southern Australia and RWA is consequently considered a major threat to cereal production in Australia as well [8]. In an updated distribution model for predicting potential spread of RWA, Avila et al. [9] suggested that RWA would be able to establish in all major wheat- and barley-growing regions in New Zealand. The first record of RWA outside its original area of distribution was in South Africa in 1978. Initially the distribution was confined to the Bethlehem area in the Eastern Free State, but by 1979, the RWA had spread to other wheat-producing areas in the country [6]. The first record of RWA in the United States was in 1986 [5]. RWA invaded all the Central European countries from the south-east [10] and was first detected in the Czech Republic in 1993 [11, 12]. It was found that RWA expanded from its Mediterranean distribution range to the northwest. It seems that the expansion route has covered Serbia, Hungary and the Czech Republic [11]. Puterka et al. [13] determined that the origin of populations distributed in South Africa, Central and North America was in Turkey with an indication of random establishment by commerce rather than through migration. Zhang et al. [14], however, found evidence of long-term existence and expansion of RWA in China and speculate that RWA are not frequently transported by human agricultural activities. With the expansion of wheat fields it is possible that aphid populations may spread to areas via natural pathways such as flight or wind currents. Once established in an area RWA is very adaptable to changes in the environment. Because of its wide distribution, considerable effort has gone into developing management strategies against this global wheat pest. Currently there are two management options: breeding for deployment of resistant wheat cultivars and chemical control.

RWA-resistant cultivars were released and deployed in South Africa during 1992, and more than 70% of the wheat production area in South Africa was planted with Russian wheat aphid-resistant cultivars [15]. The durability of resistant cultivars was, however, challenged by the occurrence of RWA biotypes, first in Colorado in 2003 [16], and in South Africa in 2006 [17]. Russian wheat aphid biotypic variation was also found in Hungary [18] and Chile [19]. Since 2006, five distinct RWA biotypes have been recorded in the wheat production areas of the Eastern Free state (summer rainfall area), South Africa, RWASA2 in 2006; RWASA3 in 2009; RWASA4 in 2011 and most recently RWASA5 in 2018.

The second management option, chemical control, is also practiced in South Africa, mainly in the Western Cape (winter rainfall area) and on irrigation wheat in central and western Free State and Northern Cape. Chemical control has long term, negative impacts on the environment, especially other insect groups such as predators, pollinators, and decomposers. Hill, et al. [20] demonstrated that broad spectrum pesticide application in grain crops can lead to secondary outbreaks of pests due to alteration of natural enemy communities. The active ingredients registered for RWA control on wheat in South Africa are limited and include acetamiprid, chlorpyrifos, chlorpyrifos + cypermethrin, demeton-S-methyl, dimethoate, imidacloprid, parathion, prothiofos and thiamethoxam. With widespread and continuous use of these active ingredients, there is the possibility that RWA can build up resistance against these specific active ingredients. About 20 species in the Aphididae have evolved resistance to insecticides [21] and can be associated with detectable changes in reproductive rates [22]. Brewer and Kaltenbach [23] demonstrated that there is detectable variation in RWA insecticide susceptibility and reproductive rates after exposure to chlorpyrifos. Chlorpyrifos selection seen in wheat production may result in large scale changes in susceptibility and control failures. Russian wheat aphid variation in virulence to small grains occurs [24, 25] as well as variation in fecundity [26, 27]. There is a possibility that RWA can also evolve virulence to active ingredients in chemicals. In their recommendations for managing RWA expansion into all major grain regions of Australia Ward et al. [28] include sustainable management practices, given the somewhat indiscriminate use of insecticides to control RWA to date. They also include regular testing of field populations for evolution of insecticide resistance in their recommendations. To determine how RWA populations change over time annual monitoring was done from 2010 to 2019 in the wheat production areas of South Africa. The most recent observations is discussed here.

2. Material and methods

2.1 Survey and collection of RWA at landscape level

RWA samples were collected annually during the wheat growing season in South Africa from 2010 to 2019. All main wheat production areas within the known distribution of the RWA were sampled. The same areas were sampled each year and where possible the same fields (**Figures 1** and **2**). There are two main dryland wheat production areas in South Africa where RWA commonly occur, the Western Cape (winter rainfall area) (**Figure 1**) and the Free State (a summer rainfall area) (**Figure 2**), with irrigated wheat production areas in the Central and Western Free State and Northern Cape (**Figure 2**). Sampling sites were selected off primary or secondary roads that transected major wheat or barley production areas. Sites were 10-20 km apart with distances depending on the continuity of wheat fields. In the Western Cape an average of 32 fields were sampled (**Figure 1**) and in the Free State an average of 61 fields were sampled (**Figure 2**). Samples were collected from cultivated wheat, barley and oats as well as volunteer wheat, wild oats, rescue grass and false barley in road reserves and around cultivated fields. Infested leaves were placed in Petri dishes containing moist filter paper and stored in an icebox for transportation to the glasshouse. The number of aphids per plant, percentage plants infested, growth stage of the plants and damage on the plants were recorded. The geographical co-ordinates and elevation where the samples were collected were also captured on a GPS and all the information of each sample collected was entered into a database (Windows Office –Excel).

2.2 Establishing clone colonies of collected RWA samples

A single female aphid from each sample collected in the field was transferred to a wheat plant and caged (gauze size: 315micron) to produce a clone colony. RWA clone colonies are kept in glasshouse cubicles at night/day temperatures of 16 °C/22 °C and maintained on various wheat cultivars to avoid pre-adaptation to a specific cultivar until they multiplied sufficiently to be used for screening. Each clone colony is cultured for an average period of two to three months before screening.

2.3 Screening of clone colonies of collected RWA samples for determination of potential biotypes

The biotype of each RWA clone was determined by screening its feeding damage on 11 previously established plant resistant sources containing designated resistance genes *Dn*1 to *Dn*9 and *Dnx* and *Dny* (**Table 1**). Infestations of RWASA1 cause susceptible damage symptoms on wheat entries containing the *Dn2* and *Dn*3 gene (**Table 1**). RWASA2 cause susceptible damage symptoms on wheat entries containing *Dn*1, *Dn2*, *Dn*3, *Dn8* and *Dn*9 resistance genes (**Table 1**). RWASA3 is distinguished from RWASA2 by its added virulence to *Dn*4 and RWASA4 is distinguished from RWASA3 by its added virulence to *Dn5* (**Table 1**). RWASA5 is distinguished from RWASA4 by its added virulence to *Dn*6 and *Dn*x (**Table 1**).

Ten seeds of each plant entry were planted in a seedling tray filled with sterilized sand in a randomized complete block design with four replications for each biotype

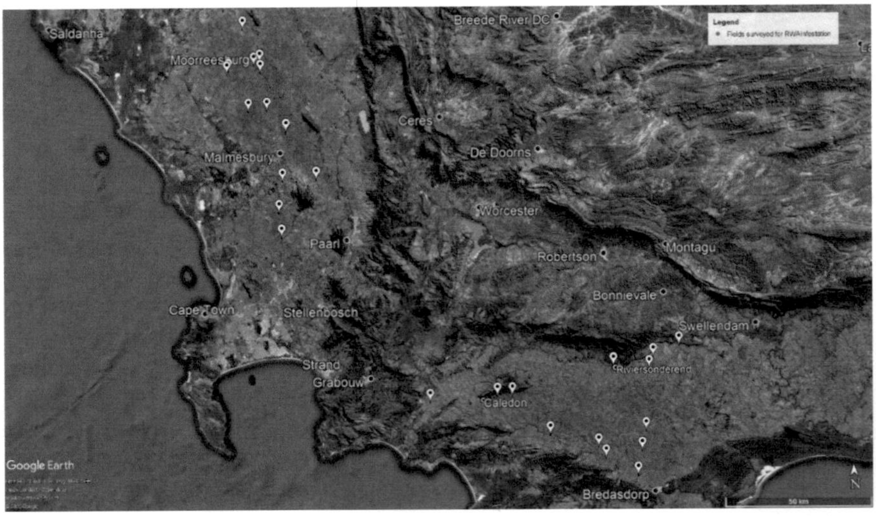

Figure 1.
Sampling sites for Russian wheat aphid (RWA) in the Western cape (winter rainfall area), South Africa from 2010 to 2019.

Figure 2.
Sampling sites for Russian wheat aphid (RWA) in the Free State (summer rainfall area), South Africa from 2010 to 2019.

determination. Plant entries were randomly assigned to rows and were separated by border rows planted with RWA susceptible Tugela. Plants were kept in glasshouse cubicles at night/day temperatures of 16 °C/22 °C, natural light. Immediately after planting, the seedling trays were placed in gauze (315micron) cages to avoid contamination by secondary aphids. Plants were infested at the two-leaf stage with collected RWA clone colonies. Plants were rated with a ten-point damage rating scale, which included leaf chlorosis and leaf rolling [29]. A score from 1–4 describes leaf chlorosis, 5–6 striping on the leaves and 7–10 rolling. Once the susceptible wheat Tugela showed susceptible damage symptoms, all plants were rated. RWA biotypes were classified

Wheat genotype	*Dn* R gene gengene	RWASA1	RWASA2	RWASA3	RWASA4	RWASA5
CO03797	*Dn1*	R	S	S	S	S
CO03804	*Dn2*	S	S	S	S	S
CO03811	*Dn3*	S	S	S	S	S
Yumar	*Dn4*	R	R	S	S	S
CO9500043	*Dn5*	R	R	R	S	S
CO960223	*Dn6*	R	R	R	R	S
94 M370	*Dn7*	R	R	R	R	R
Karee-*Dn8*	*Dn8*	R	S	S	S	S
Betta-*Dn9*	*Dn9*	R	S	S	S	S
PI586955	*Dnx*	R	R	R	R	S
Stanton	*Dny*	R	R	S	S	S

Table 1.
Comparison of plant reaction of the five Diuraphis noxia *biotypes identified in South Africa.*

by using damage ratings for each plant entry where the plant was considered resistant (R) if the damage rating was 1–6.5 and susceptible (S) if the damage rating was above 6.5–10. Each clone was given a biotype designation based on the differential virulence profile to the *Dn*1 to *Dn9* and Dnx and Dny resistance genes (**Table 1**).

Biotype (clones) groups across all plant differentials were analyzed using a two-way (clone, plant entry) analysis of variance (ANOVA). Mean damage rate entries with significant ($P < 0.05$) clone-by-plant interactions were separated by Fisher's protected least significant difference (LSD) test at the 5% level (SAS Institute 2003).

3. Results and discussion

Representative samples of five RWA biotypes were collected in the different wheat production areas in South Africa, with a range of different climatic conditions and different host plants from 2010 to 2019 (**Figures 1** and **2**). The number of samples collected in a specific area varied depending on the area planted with wheat or barley or the availability of alternative hosts and the level of infestation. An average of 32 fields were sampled in the Western Cape (**Figure 1**) and 61 in the Free State (**Figure 2**). Environmental conditions, including temperature, humidity, rainfall, soil type and availability of host plants play an important role in the population increase and distribution of different RWA biotypes. Because these variables change from year to year and between different areas, the distribution of RWA biotypes will vary over years and between different geographical areas.

Analysis of the main effects of damage rating for the five Russian wheat aphid biotype colonies indicated a significant clone ($F = 117.48$; $df = 3$; $P < 0.0001$), plant entry ($F = 133.59$; $df = 11$; $P < 0.0001$) and clone-by-plant entry interaction ($F = 12.82$; $df = 33$; $P < 0.0001$), suggesting that the plant entries responded differently to the different aphid clones. Biotypes are identified by the distinct feeding damage responses they produce on wheat carrying different RWA resistance genes from *Dn*1 to *Dn*9 [30]. Infestations of RWASA1 caused susceptible damage symptoms on the wheat entry containing the *Dn2* and *Dn*3 gene (**Table 1**). RWASA2 caused susceptible damage symptoms on wheat entries containing *Dn*1, *Dn2*, *Dn*3,

Dn8 and *Dn9* resistance genes (**Table 1**). RWASA3 is distinguished from RWASA2 by its added virulence to *Dn4* and RWASA4 is distinguished from RWASA3 by its added virulence to *Dn5* (**Table 1**). RWASA5 was the most virulent biotype in South Africa with susceptible responses to ten plant differentials containing ten different *Dn* genes (**Table 1**). Randolph et al. [31] found the American RWA2 to be the most virulent strain tested with susceptible responses to 12 plant differentials.

The concentration of RWA biotypes occurred mainly in the Eastern Free State with very few wheat fields infested with RWASA1 (original biotype, reported in 1978). RWASA1 occurred mainly in the Western Free State and Northern Cape. Since 2006, five distinct RWA biotypes have been recorded in the wheat production areas of the Eastern Free State, RWASA2 in 2006; RWASA3 in 2009; RWASA4 in 2011 and RWASA5 in 2018. The populations of RWA biotypes fluctuated over the years with RWASA2 being the dominant biotype from 2010 to 2011, RWASA3 dominating from 2012 to 2013 and RWASA4 from 2014 to 2016 (**Figure 3**). During the 2018 season RWASA5, was recorded for the first time on 8 wheat fields in the Lindley, Reitz and Danielsrus areas in the Eastern Free State. During 2019 this biotype had increased and spread to other areas of the Eastern Free State and was recorded on 12 wheat fields in the Eastern Free State. This biotype was dominant from 2018 to 2019 (**Figure 3**). Merrill et al. [32] found, in a general survey of aphid mixtures for virulence to resistant Yumar (with *Dn4* gene) in Colorado from 2004 to 2008, that *Dn4* virulence increased from 82% in 2005 to 98% in 2008. When a new RWA biotype appear, this new biotype seem to be able to outcompete the previous biotypes in the area and displace the other biotypes. Puterka et al. [33] found, in an area-wide study in the USA during 2005, that RWA2 almost completely displaced the original biotype. A survey from 2010 to 2013 revealed a change in biotypic diversity of RWA populations in the United States, with RWA 1,6 and 8 across regions showing high percentages during 2011 (64–80%) and 2013 (69–90%) [34]. In South Africa RWA biotype with added virulence to genes used in resistant wheat cultivars were recorded every 2 to 3 years in the Eastern Free State where RWA resistant wheat cultivars were commonly deployed. These newly recorded RWA biotypes became the dominant biotype in these areas until a more virulent RWA biotype was recorded (**Figure 3**). The most recently recorded biotype during 2018, RWASA5, is virulent against all known *Dn* genes used in wheat except *Dn7* (94 M370) (**Table 1**).

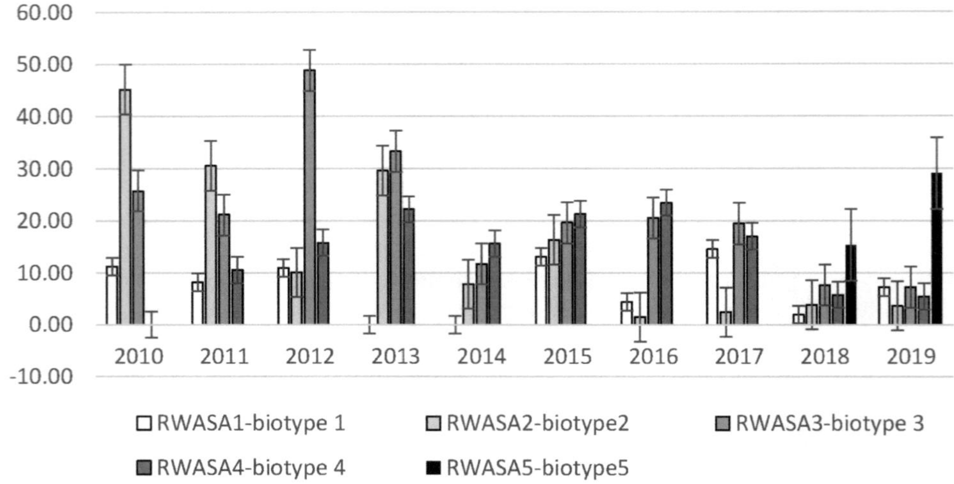

Figure 3.
Russian wheat aphid (RWA) SA biotype distribution in the Free State, South Africa (summer rainfall area) from 2010 to 2019 (average fields sampled: 61).

With the increase and spread of more virulent RWA biotypes the use of insecticides may again become the main management option in these areas. Merrill et al. [35] found that even though resistant wheat cultivars historically provided excellent management of RWA on wheat crops in Colorado, the increase of new RWA biotypes resulted in all commercially available winter wheat cultivars being susceptible to RWA feeding damage and associated yield losses. This led to insecticides once again becoming the main management tactic used on Colorado wheat [35]. In the Western Cape, where chemical control is the most common control measure for RWA, RWASA1 remained the only biotype and the biotype diversity seen in the Eastern Free State was not experienced in this area. There was however, an increase in RWASA1 incidence in the Western Cape from between 30 to 60% fields infested from 2010 to 2016 to between 70 to 100% fields infested with RWA from 2017 to 2019 on the fields that were annually surveyed (**Figure 4**). In a survey of farmers in the Western Cape during the 2017 wheat production season 75% of the respondents observed RWA on their crops [36]. All these farmers use chemical control, in the form of preventative spray, to control RWA, because it is cheap and effective [36]. The fact that RWASA1 became more widespread in the Western Cape and that in some cases live populations were collected in fields recently sprayed with insecticides may indicate insecticide resistance. The active ingredients registered for RWA control on wheat in South Africa are limited and include acetamiprid, chlorpyrifos, chlorpyrifos + cypermethrin, demeton-S-methyl, dimethoate, imidacloprid, parathion, prothiofos and thiamethoxam. The most common active ingredients used by producers in the Western Cape are chlorpyrifos, dimethoate, imidacloprid and thiametoxam (Mr K. Naicker, Cape RnD, Meridian Agritech). In the Western USA, chlorpyrifos was the predominantly used insecticide, with area-wide treatment of wheat acreage in specific localities [37]. Puterka et al. [13] detected genetic variation and potential for biotypic diversity in RWA among world-wide collections of RWA from countries in Eurasia, South Africa and the United States in 1990. This variation in other traits may be indictors of adaptations, which could confer RWA resistance to chlorpyrifos [23]. Brewer and Kaltenbach [23] demonstrated that variation in RWA susceptibility to chlorpyrifos and associated reproductive rates occur in the small grains growing region of the USA. Furthermore, approximately 20 species in the Aphididae have evolved resistance to insecticides [21] that can be

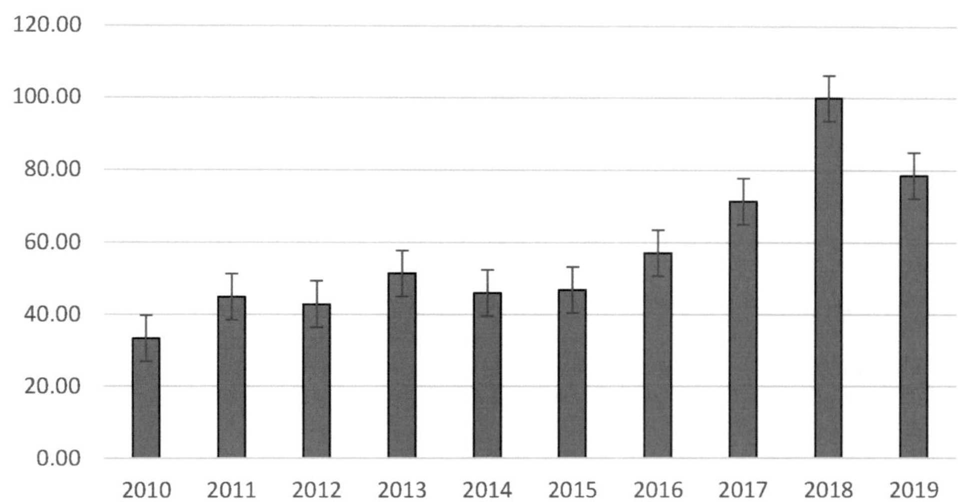

Figure 4.
Percentage of wheat fields surveyed in the Western cape, (winter rainfall area), South Africa, infested with Russian wheat aphid (RWA) SA biotype1 (average fields sampled: 32).

associated with detectable changes in reproductive rates [22]. In South Africa RWA showed considerable biotypic adaptation and change in reproductive rate to resistant wheat [25, 27, 38], resulting in five RWA biotypes occurring in wheat production areas where RWA resistant wheat were deployed in the Eastern Free State. This may be an indication that RWA in South Africa have the adaptive ability to develop resistance to active ingredients of insecticides used to control them in the Western Cape. Large-scale changes in susceptibility were detected in other aphids in which consistent and severe selection pressure occurred [21]. Brewer and Kaltenbach [23] stated that even though control failure problems have not been reported, periodic assessment of RWA populations of field derivation is necessary. Ward et al. [28] also recommend regular testing of field populations to understand if insecticide resistance is likely to evolve in Australia. According to Brewer and Elliott [39] better understanding of the mediating effects of host plant and habitat manipulations may accelerate our ability to plan cereal production systems with improved ability to suppress cereal aphids, including future invading species.

4. Conclusion

Given the invasive ability, evolutionary adaptability to changing conditions, virulence, and fecundity of RWA, it remains a threat to global wheat production and wheat cultivation. RWA remain present in all the wheat production areas of South Africa and these populations are becoming more virulent as indicated by the spread of the recently recorded biotype, RWASA5, in the Eastern Free State. Management practices in different regions of South Africa may cause increased virulence in RWA populations. Based on these observations testing of field populations to understand if insecticide resistance is evolving in RWA populations in the Western Cape is warranted. It is important that future management practices focus on sustainability instead of the indiscriminate use of insecticides globally to control RWA to date. Increasing diversity in fields through undersowing, reduced tillage, intercropping and incorporation of cover crops will be an effective start to sustainable management practices. Vegetation strips have favorable microclimate for survival of generalist predators, and alternative prey and resources during winter, resulting in higher densities of generalist predators in cereal fields [40, 41]. This together with minimal use of insecticides, only when necessary, will increase the insects providing ecosystem services and predators, parasitoids and pathogens that will keep RWA populations and economical damage low. Management approaches against cereal aphid invasions differ depending on aphid ecology, specific system influences, and local management practices [42]. Any practice based on aphid population monitoring that facilitates threshold-based insecticide use will be effective across agroecosystems, with area-wide management systems being most appropriate to large-scale cereal production systems.

Acknowledgements

The author wish to acknowledge the Agricultural Research Council (ARC)-Small Grain for facilities provided and the Winter Cereal Trust (WCT) for research funding.

Conflict of interest

The author declare no conflict of interest.

Author details

Astrid Jankielsohn
Agricultural Research Council (ARC)-Small Grain, Bethlehem, South Africa

*Address all correspondence to: jankielsohna@arc.agric.za

References

[1] Colautti, R.I. and MacIsaac H.J. 2004. A neutral terminology do define 'invasive' species. Diversity and Distributions. 10: 135-141.

[2] Prentis, P.J., JRU. Wilson, E.E. Dormontt, D.M. Richardson, and A.J. Lowe. 2008. Adaptive evolution in invasive species. Trends Plant Sci. 13: 288-294.

[3] Liu, X., J.L. Marshall, P. Stary, O. Edwards, G. Puterka, L. Dolatti, M. El Bouhssini, J. Malinga, J. Lage, and C.M Smith. 2010. Global phylogenetics of Diuraphis noxia (Hemiptera: Aphididae), an invasive aphid species: evidence for multiple invasions into North America. J. Econ. Entomol. 103(3): 958-965. DOI: 10.1603/EC09376

[4] Durr, H.J.R. 1983. *Diuraphis noxia* (Mordvilko) (Hemiptera: Aphididae), a recent addition to the aphid fauna of South Africa. Phytophylactica 15: 81-83.

[5] Morrison, W. P., and F. B. Peairs. 1998. Response model concept and economic impact, pp. 1-11. In S. S. Quisenberry and F. B. Peairs (eds.), A response model for and introduced pest – the Russian wheat aphid (Homoptera: Aphididae). Thomas Say Publications in Entomology, Entomological Society of America, Lanham, MD.

[6] Walters, M.C, F. Penn, F. Du Toit, T.C. Botha, K. Aalbersberg, P.H. Hewitt & S.W. Broodryk. 1980. The Russian wheat aphid. Fmg S. Afr. Leaflet Series Wheat G3, 1-6

[7] Havelka, J., M. _Zurovcov_a, S. Rychly´, and P. Stary´. 2013. Russian wheat aphid, *Diuraphis noxia* in the Czech Republic – cause of the significant population decrease. J. Appl. Entomol. 138: 273-280.

[8] Yazdani, M., Baker, G., DeGraaf, H., Henry, K., Hill, K., Kimber, B., Malipatil, M., Perry, K., Valenzuela, I. and Nash, M.A. 2017. First detection of Russian wheat aphid *Diuraphis noxia* Kurdjumov (Hemiptera: Aphididae) in Australia: a major threat to cereal production. Austral Entomology. doi: 10.1111/aen.12292

[9] Avila, G.A., Davidson, M. van Helden, M. and Fagan, L. 2019. The potential distribution of the Russian wheat aphid (*Diuraphis noxia*): and updated distribution model including irrigation improves model fit for predicting potential spread. Bulletin of Entomological Research. 109: 90-101.

[10] Basky, Z., and V. F. Eastop. 1989. *Diuraphis noxia* in Hungary. Newslett. Barley Yellow Dwarf 4: 34.

[11] Stary´, P. 1996. The expansive Russian wheat aphid Diuraphis noxia (Kurdj.) detected in the Czech Republic. J. Pest Sci. 69: 19-20.

[12] Vosˇlajer, Z. 1999. *Diuraphis noxia* in the Czech Republic and relationship between the migration and the temperature conditions, pp. 589-596. In Proceedings, 5th International Conference on Pests in Agriculture, Part 2. Montpellier, France.

[13] Puterka, G.J., Black, W.C., Steiner, W.M., Burton, R.L. 1993. Genetic variation and phylogenetic relationships among worldwide collections of the Russian wheat aphid *Diuraphis noxia* (Mordvilko), inferred from allozyme and RAPD-PCR markers. Heredity. 70: 604-618.

[14] Zhang, B., Edwards, O.R., Kang, L. and Fuller, S.J. 2012. Russian wheat aphids (*Diuraphis noxia*) in China: native range expansion or recent introduction? Molecular Ecology 21: 2130-2144.

[15] Marasas, C., P. Anandajayasekeram, V.L. Tolmay, P. Martella, J.L. Purchace,

and G.J. Prinsloo. Socio-economic impact of Russian wheat aphid control research program. SACCAR/ ARC Report, SACCAR, Gaborone, Botswana, (1999). 147pp. http:// agris.fao.org/agris-search/search. do?recordID=QY2001000257

[16] Haley, S.D., F.B. Peairs, C.B. Walker, J.B. Rudolph, and T. L. Randolph. 2004. Occurrence of a new Russian wheat aphid biotype in Colorado. Crop Sci. 44: 1589-1592.

[17] Tolmay, V.L., R.C. Lindeque & G.J. Prinsloo. 2007. Preliminary evidence of a resistance-breaking biotype of the Russian wheat aphid, *Diuraphis noxia* (Kurdjumov) (Homoptera: Aphididae), in South Africa. African Entomology 15(1): 228-230

[18] Basky, Z. 2003. Biotypic and pest status differences between Hungarian and South African populations of Russian wheat aphid, *Diuraphis noxia* (Kurdjumov) (Homoptera: Aphididae). Pest Manag. Sci. 59: 1152-1158.

[19] Smith, C.M., T. Belay, C. Stauffer, P. Stary, I. Kubeckova, and S. Starkey. 2004. Identification of Russian wheat aphid (Homoptera: Aphididae) populations virulent to the *Dn*4 resistance gene. J. Econ. Entomol. 97: 1112-1117.

[20] Hill, M.P., Macfadyen, S., and Nash, M.A. 2017. Broad spectrum pesticide application alters natural enemy communities and may facilitate secondary pest outbreaks. PeerJ 5:e4179; DOI 10.7717/peerj.4179

[21] Devonshire, A.L. 1989. Occurrence of insecticide-resistant aphids. In A.K Minks and P. Harrewijn (eds.), Aphids, their Biology, Natural Enemies and Control, Vol. C, pp. 123-139. Elsevier, NY.

[22] O'Brien, P.J. and J.B. Graves. 1992. Insecticide resistance and reproductive potential of *Aphis gossypii* Glover. Southwestern Entomol. 17:115-121.

[23] Brewer, M.J. and Kaltenbach, J.E. 1995. Russian wheat aphid (Homoptera: Aphididae) population variation in response to chlorpyrifos exposure. Journal of the Kansas Entomological Society. 68(3): 346-354.

[24] Puterka, G.J., J.D. Burd, and R.L. Burton. 1992. Biotypic variation in a world-wide collection of Russian wheat aphid (Homoptera: Aphididae). J. Econ. Entomol. 85: 1497-1506.

[25] Jankielsohn, A. 2011. Distribution and Diversity of Russian Wheat Aphid (Hemiptera: Aphididae) Biotypes in South Africa and Lesotho. J. Econ. Entomol. 104(5): 1736-1741.

[26] Webster, J.A., F. Du Toit, and T.W. Popham. 1993. Fecundity comparisons of the Russian wheat aphid (Homoptera: Aphididae) in Bethlehem, South Africa, and in Stillwater, Oklahoma. J. Econ. Entomol. 86: 544-548.

[27] Jankielsohn, A. 2013. Host Associations of *Diuraphis noxia* (Homoptera: Aphididae) Biotypes in South Africa. Journal of Economic Entomology 06(6):2595-2601.

[28] Ward, S., Van Helden, M., Heddle, T., Ridland, P.M., Pirtle, E. and Umina, P.A. 2020. Biology, ecology and management of Diuraphis noxia (Hemiptera: Aphididae) in Australia. Austral Entomology. 59(2): 238-252. https://doi.org/10.1111/aen.12453

[29] Tolmay, V. L. 1995. The inheritance and mechanisms of Russian wheat aphid (*Diuraphis noxia*) resistance in two *Triticum aestivum* lines. M.Sc. thesis, University of the Orange Free State, Bloemfontein, South Africa.

[30] Puterka, G.J., Nicholson, S.J., Brown, M.J., Cooper, W.R., Peairs, F.B. and Randolph, T.L. 2014.

Characterization of eight Russian wheat aphid (Hemiptera: Aphididae) biotypes using two-category resistant-susceptible plant responses. J. Econ. Entomol. 107(3): 1274-1283. DOI:http//dx.doi.org/10.1603/EC13408

[31] Randolph, T.L., F. Peairs, A. Weiland, J.B. Rudolph, and G. J. Puterka. 2009. Plant responses to seven Russian wheat aphid (Hemiptera: Aphididae) biotypes found in the United States. J. Econ. Entomol. 102 (5): 1954-1959.

[32] Merrill, S.C.,Walker, C.B., Peairs, F.B., Randolph, T.L., Hayley, S.D., and Hammon, R.W. 2009. Displacement of Russian wheat aphid, Diuraphis noxia (Kurdjumov), Biotype 1 in Colorado by Russian what aphid biotypes virulent to the wheat resistance gene Dn4. Colorado State University Experiment Station Technical Bulletin TB09-09.

[33] Puterka, G.J., Burd, J.D., Porter, D., Shufran, K., Baker, C., Bowling, B., and Patrick, C. 2007. Distribution and diversity of Russian wheat aphid (Hemiptera: Aphididae) in North America. J. Econ. Entomol. 100: 1679-1684. https://doi.org/10.1093/jee/100.5.1679.

[34] Puterka, G.J., Giles, K.L., Brown, M.J., Nicolson, J., Hammon, R.W., Peairs, F.B., Randolph, T.L., Michaels, G.J., Bynum, E.D., Springer, T.L., Armstrong, J.S. and Mornhinweg, D.W. 2015. Change in biotypic diversity of Russian wheat aphid (Hemiptera: Aphididae) populations in the United States. J. Econ. Entomol. 1-7. DOI:10.1093/jee/tov008.

[35] Merrill, S.C., Holtzer, T.O., Peairs, F.B. and Lester, P.J. 2009. Modeling sparial variation of Russian wheat aphid overwintering population densities in Colorado winter wheat. J. Econ. Entomol. 102(2): 533-541.

[36] De Lange, W. 2017. Monetary valuation of the impact of aphids on selected commercial small grains in the Western Cape. CSIR Natural Resources and the Environment, July 2017.

[37] Legg, D.E., Ferrell, M., Taylor, D.T. and Kellogg, D.L. 1992. Pesticide use in Wyoming, 1990. RJ-211, Coop. Ext. Serv., Univ. of Wyoming, Laramie, Wyoming.

[38] Jankielsohn, A. (2019). New Russian wheat aphid biotype found in the Free State. SA Grain, March 2019. P 70-71.

[39] Brewer, M.J. and Elliott, N.C. 2004. Biological control of cereal aphids in North America and mediating effects of host plant and habitat manipulation. Annu. Rev. Entomol. 49: 219-42. Doi:10.1146/annurev.ento.49.061802.123149.

[40] Sunderland, K. and Samu, F. 2000. Effects of agricultural diversification on the abundance, distribution, and pest control potential of spiders: a review. Entomol. Exp. Appl. 95: 1-13.

[41] Jmhasly, P. and Nentwig, W. 1995. Habitat management in winter wheat and evaluation of subsequent spider predation on insect pests. Acta Oecol. 16: 389-403.

[42] Brewer, M.J., Peairs, F.B., and Elliott, N.C. 2019. Invasive cereal aphids of North America: ecology and pest management. Annual Review of Entomology. 64:73-93. https://doi.org/10.1146/annurev-ento-011118-111838

Chapter 6

Silicon Use in the Integrated Disease Management of Wheat: Current Knowledge

Leandro José Dallagnol, Andrea Elizabeth Román Ramos and Keilor da Rosa Dorneles

Abstract

Silicon (Si) is a benefic element for higher plants such as wheat (*Triticum aestivum*) in which it is accumulated in the shoot tissues. In this crop, leaf diseases and spike diseases are the cause of yield losses, and therefore several studies had been conducted under field and greenhouse conditions to demonstrate that plants supplied with Si reduced most of the diseases damage due to the amelioration of the plant defenses. However, the benefits of Si depend on its accumulation in the plant's tissue, which is influenced by the availability of the element in the soil as well as the up-take ability of the wheat cultivar. In this chapter we present the current knowledge about the mechanisms of Si absorption and its accumulation in different tissues of the wheat plant, the most studied options for silicate fertilization, and the benefits of Si on grain yield. We also present some insight of the effect of Si-supply in wheat on the reduction of main leaf and ear diseases, bringing evidence and explanation of the defense mechanisms involved. In addition, we provide an overview of the Si effect on the physiology (gas exchange, chlorophyll *a* fluorescence and carbohydrate metabolism) of the wheat plant. Finally, questions have been raised about the Si uses as fertilizer that still needs to be answered. We recognized that some studies have enhanced our understanding of Si providing evidence of the Si use as disease management strategy, but further research is needed to make the Si uses a simple task for wheat growers under field condition.

Keywords: disease control, diseases management, silicate fertilization, sustainable management, wheat diseases, wheat yield

1. Introduction

Wheat (*Triticum aestivum* L.) is the most cultivated crop on Earth [1, 2] being a key cereal for global food security. Wheat provides calories to 85% of the world population (contributing of 5 to 57% of daily consumed calories, depending of the country) and proteins to more than 82% of the world population (contributing of 6 to 60% of daily calories intake, depending of the country) [3]. Historically (1961–2009) the increase in the world production of wheat occurred primarily due to the increase in productivity which supplied the increase in the demand for this cereal [4]. As the global population continues to increase, the world demand for wheat is predicted to continue raising [2, 4], being forecast that in order to feed the world population in 2050 it will be necessary to almost double the current wheat production [5]. The challenges to achieving this production target include abiotic (drought, heat and salinity) and biotic (insects, pathogens and weeds) stresses that can be enhanced by climate changes [5].

Pathogens are among the main threats to high yield of wheat and a threatening to food security. Wheat is affected by many pathogens, however their occurrence and yield loss, estimated for each disease, vary from country to country and season to season. The main aboveground wheat diseases worldwide are rusts (*Puccinia* spp.), septoria nodorum blotch (*Parastagonospora nodorum*), septoria tritici blotch (*Zymoseptoria tritici*), tan spot (*Pyrenophora tritici-repentis*), fusarium head blight (*Fusarium graminearum* species complex), powdery mildew (*Blumeria graminis* f. sp. *tritici*) and wheat blast (*Magnaporthe oryzae* Pathotype Triticum).

The control of these diseases is carried out preferably through resistant cultivars. However, for some of the diseases, there are no cultivars with sufficient resistance to contain the damage in yield or the resistance is ephemeral, especially when governed by race-specific resistance genes, due to the rapid evolution of the pathogen [6–9]. As a result, the use of fungicides is common in wheat crops, but it raises the cost of production and it does not always give the expected control for some diseases [10, 11]; furthermore there is risk of development of resistance in the fungal to fungicides [12].

In this scenario, silicon (Si) become as an attractive alternative to be included in the management of wheat crop. Silicon is a mineral element considered benefic to plants, however in many soils its concentration available for plants is low [13]. In these soils, fertilization with Si sources has shown positive results. Numerous studies demonstrate the beneficial effect of Si in relieving abiotic stresses and in the control of biotic diseases on Si-accumulator plants (reviewed by [14–17]). This chapter presents the current knowledge on Si up take by the wheat plant, its effects on grain productivity and wheat technological quality, physiological aspects, and biochemical and histological defenses enhanced by the element, on several wheat-pathogen interaction.

2. Silicon wheat absorption

The knowledge of Si absorption has been studied in different plants such as monocotyledonous and dicotyledonous species providing evidence to explain the process [18–21]. Initially was believed that the transpiration was the main factor determining Si uptake in plants. New evidence confirmed that the Si absorption and accumulation could be explained by the active transport mechanisms inherent to the roots and the shoots.

In wheat, the first evidences of active transport mechanisms come with studies showing that approximately 90% of the Si absorbed by the plant was transferred to the shoots, maintaining the roots in a relatively low-Si status [22, 23]. Later, Mayland et al. [24] reported that the amount of Si accumulated by the wheat plant was higher than expected to occur only via transpiration providing data to support the classification of wheat as a Si accumulator (accumulating Si in concentration up to 20 g kg^{-1} of dry weight). Advancing, Rafi and Epstein [25] reported that Si is rapidly absorbed by wheat plants from solution containing Si at 0.5 mM, a concentration near of that of the element in soil solutions, and the uptake rate were similar between plants 'preloaded' with Si and plants grown previously in solutions without Si addition. Further studies demonstrated that Si uptake by wheat is under metabolic control due to the absorption of Si show a concentration dependence obeying Michaelis–Menten kinetics and it is affected by metabolic inhibitors (dinitrophenol and potassium cyanide) [26]. Later, Montpetit et al. [21] cloned and functional characterized the *TaLsi1*, a wheat Si transporter gene, which is an ortholog of *OsLsi1* from rice. The genes *TaLsi1* and *OsLsi1* belong to Nod26-like intrinsic proteins (NIPs) III subgroup of the aquaporin membrane protein family.

Thus, the Si absorption is facilitated by specific NIPs with a distinct selectivity that facilitate the passive transport of water and/or small uncharged solutes such as monosilicic acid [$Si(OH)_4$] [27]. According to Ma and Yamaji [19] specific NIPs as *Lsi1* (Si influx transporter) facilitates the passive transport of Si across the plasma membrane from the environment (external solution) to the plant cell in the form of [$Si(OH)_4$], and efflux transporters known as *Lsi2* mediate the loading of Si into the xylem to facilitate root-to-shoot translocation, which, in turn, moves Si to the aerial parts where it is deposit as amorphous Si (SiO_2). According to the authors, these Si transporters are localized to the plasma membrane, but, in different plant species, show different expression patterns and tissue or cellular localizations that are associated with different levels of Si accumulation [19]. In this context, the molecular characterization and phylogeny of the Si permeable channel, *TaLsi1*, which is expressed only in the roots and independent from Si concentrations, can explain the Si absorption by wheat plants [14, 21].

The concentration of Si on wheat tissue varies according to the soil and cultivar. For example, a study conducted in two locations (Abed and Sejet, Denmark) showed that Si concentration in the wheat straw ranged from 11.3 g kg^{-1} to 23.4 g kg^{-1} of dry weight. The study performed with 20 genotypes, showed that on average wheat grown in Abed contained 25% more Si than wheat grown in Sejet, which as attributed to variation in edaphic factors such as soil pH or silicate mineral composition which affect the Si availability to plants [28]. In regarding to wheat genotypes, the difference between the lowest and highest Si concentration was 75% at Abed and 44% at Sejet reflecting differences in the ability of roots to take up Si from the soil solution [28]. Carter et al. [29] and Ranjbar et al. [30] also observed difference among wheat cultivars regarding Si concentration in the shoot. Ranjbar et al. [30] also showed that there is a relationship between shoot Si concentration and Si acquisition efficiency. These studies clearly indicated that accumulation of Si in the wheat shoot is variable among cultivars which may influenced by the cultivar ability to absorption and also by the availability of Si in the soil.

After uptake by roots, Si follow the transpiration flow and it is accumulated beneath cuticle forming a double layer Si-cuticle, associated to cell wall and in Si-accumulating cells [31]. Furthermore, it was been previously reported that the highest Si concentration was present in major transpiration parts of the plants followed by the other parts of the plants [32]. In wheat, the highest silicified cells were present in leaf blade followed by the awn, leaf sheath, lemma, rachilla and stem, thus leaf blade contains the highest Si concentration [33]. In agreement, another study showed that the accumulation of Si was highest in vegetative tissue (leaf blades > leaf sheaths > stem) and lowest in grain followed by roots, increasing with increasing stomata density in the tissues [34]. In awns, the number of silicified cells was linearly correlated to Si concentration in dry weight which suggests cellular control over silicification [35]. Using scanning electron microscopic, authors found a continuous silica layer under the cuticle, extended silicification in the epidermis cell wall and in sclerenchyma cells near the vascular bundles, but not in the stomata, suggesting that an active process directs the soluble Si away from the water evaporation stream [35]. On the leaves, X-ray microanalysis revealed that Si was deposited in a linear pattern that corresponded to the silica cells, being greater the amounts of Si in the linear areas of silica cells from plants grown in soil supplied with silicate fertilizer [36, 37]. Another study showed that Si was predominantly deposited in the epidermis cells of the leaves and their cell walls [38].

As wheat is a Si-accumulating species, it may remove considerable amount of Si if straw is removed from the field. In this context, a study considering long-term cultivated field analyzed the impact of Si accumulating plants on the biogeochemical cycle of Si and indicated that the concentration of amorphous silica is lower in cultivated soils compared to natural ecosystems, due to the amorphous Si pool decreases with time particularly in surface soil, contrary to natural ecosystems [39]. For instance, an estimation of shoot Si uptake by wheat based on 10-year average of harvested area, production level, reported biomass/harvested portion ratio and shoot Si content in United States indicated that the annual shoot Si uptake of wheat is 2.144,278 tons and 108 kg ha^{-1} [40]. In this sense, in crop systems in which the straw is removed from the field, the available Si in soils do not sustain high Si concentrations not only for wheat but also other crops in the long term [39]. Under this condition, it is clear the concern on the Si reduction from field pointing out the necessity of Si sources as fertilizers and eventually the management of the wheat straw to obtain the benefits of Si to wheat plants.

2.1 Silicon fertilization

Orthosilicic is the second most abundant element in the earths crust and plays a number of important roles in the plants. The silicic acid is present in the soil as an uncharged monomeric molecule below pH 9 [19]; their concentration in soil varying between 0.1 to 0.6 mM [41]. In the past 20 years, the scientific documentation on the benefits of Si to crops has helped establish Si fertilization as an agronomic practice in many agricultural lands worldwide [40]. Thus, it is recognized that Si fertilization confers benefits to wheat crop.

In this context, the most common Si fertilizers are wollastonite and slag (calcium silicate). In the case of wollastonite which is a natural calcium silicate [42], that contains higher fractions of easily soluble Si compared to slags [40]. It is considered to be the most efficient Si fertilizer for soil application due to that it can release the largest amount of plant available Si (2.31–3.6%) into soil solution [43, 44]; however, its use is often limited because of its relatively high cost [45]. Calcium silicate slags are by-products of the metallurgical smelting process, contain varying percentages of Si [46], and have been observed positive effects on correcting soil acidity [47], plant growth and alleviation of stresses [48–50]. Other commonly used Si fertilizers are sodium metasilicate and potassium silicate. These Si fertilizers have been found very helpful in improving growth parameters in biotic, drought and salt stress in wheat [38, 51].

In the case of pyrolitic fine silica particles, sodium metasilicate or silica gel is used for agricultural purposes. In soil, wheat plants grown under identical growing conditions, the efficiency of the Si compounds to increase the Si concentration on the plants increased in the order sodium metasilicate > silica gel >pyrolitic fine silica particles and seemed to correlate with the ease of formation of orthosilicic acid from these compounds [38]. For instance, the application of liquid and powder silicate fertilizers in the soil contributed similarly to the concentration of Si to the soil solution and doubled the Si concentration on wheat tissue [52]. Furthermore, Si uptake by wheat plant as well as its growth is significantly affected by the type of Si pool in the soil and factors controlling its solubility [53].

On the other hand, foliar application, mainly as sodium metasilicate and potassium silicate, is cause of debate due to the major portion of the Si uptake come from to the roots; however, some effects under biotic and abiotic stress have been observed (see below).

The demand of Si fertilizer due to the necessity in different agricultural environments allow the introduction and application of nano-Si fertilizer with some kind of efficient. The nano-Si is high bioavailability as smaller particle size that can be rapidly and completely form to absorb by plants and form a thick silicated layer on leaf surface [54]. In this case a study showed that addition of potassium silicate or nano-Si fertilizer in a Calcaric Cambisols increased the concentration of Si in wheat tissue, mainly in the shoot, but there is a significant relationship between the Si level/source and wheat cultivars [30].

3. Silicon and wheat yield

Silicon fertilization in the soil resulted in positive effect on grain yield and its quality, mainly under stress. In China, a four-year field experiment in Calcareous Paddy soils indicated that Si fertilization increased the wheat yield by 4.1 to 9.3% under biotic stress [55], while other studies obtained increased in the grain yield due to silicate slag fertilization ranging from 5 to 12% [55, 56].

In New Jersey, in a three consecutive years of field experiment, calcium silicate (steel slag by-product) was added on a Quakertown Silt Loam soil increasing yield up to 10%, but only under biotic stress imposed by powdery mildew [57]. A two consecutive growing seasons experiment performed in Idaho evaluated the application of Si in the form of amorphous volcanic tuff in the Greenleaf-Owyhee Silt Loam soil indicated that there was no significant effect of Si on plant height, nutrient uptake, grain yield and grain protein content of winter wheat grown in non-stressed conditions [58]. A three site-years experiment was conducted on the Alluvial Floodplain soils in Louisiana to evaluated silicate slag applications on productivity of wheat under sufficient and high nitrogen application rates showed a numerical trends of grain yield increase increasing silicate slag rate, but significant increase was only observed in one site and year [59]. According authors, the inconsistencies observed in responses to Si treatments could be due to varying physicochemical properties of soils and more research is need to better understand the effect of silicate slag use in wheat production in Louisiana.

In Brazil, calcium silicate was used as a source of soluble Si in a three-years field experiment to control shoot diseases. The results showed that wheat plants grown in soil fertilized with calcium silicate that received one application of fungicide at the stem elongation stage showed a reduction on the biotic stress increasing grain yield by 1.0 t ha^{-1} (Pazdiora, P. C. – unpublished data). Grains from these experiments was used to determine the wheat technological quality through physicochemical and rheological analyses. The data indicated that calcium silicate showed little effect on the wheat technological quality under lower disease intensity, but under higher disease intensity, it ameliorated the damage caused, keeping the technological quality near the expected level of each cultivar (Dallagnol, L. J. – unpublished data). Pot experiment evaluating three soils (Rhodic Acrudox, Rhodic Hapludox and Arenic Hapludult) indicated that application of calcium/magnesium silicate in an acid clayey Rhodic Hapludox improves the development and yield of wheat, but the silicate application in soil with pH higher to 5.3 and high Si availability does not affect the agronomic characteristics and grain yield of wheat [60].

In Poland, a two-year field experiment evaluated different methods of application of powder (diatomaceous earth) and liquid (solution of monosilicic acid) forms of Si to soil, leaves and combined methods of application (to soil and leaves) on growth parameters and yielding [61]. Authors observed that the most efficient form of Si was a liquid formulation, while powder was less effective and only in combined application achieved similar effects such as liquid Si, increasing the number of seedling emergence, the height of plants and density of spikes and yield. Furthermore, according authors, soil and foliar Si application is more effective than soil or foliar application [61]. In Germany, an experiment performed in substrate showed that Si applied in the form of an engineered nanomaterial (amorphous pyrogenic hydrophilic SiO_2) was readily taken up by the wheat plants increasing the aboveground biomass production at low (1 g SiO_2 pot^{-1}) to medium (10 g SiO_2 pot^{-1}) supply levels of Si; and grain yield at medium Si supply, probably due to increased plant phosphorus availability and nutrition [34].

Foliar Si treatment also provided some effect on wheat growth and/or yield. In Canada, foliar application of potassium silicate increased the high of wheat plants, compared to control plants, but only under biotic stress and variable according Si-based product [62]. A study performed in Iran, under greenhouse, showed that wheat plant grown in pots that received foliar application of 6 mM sodium metasilicate significantly increased biomass and grain yield, being the highest positive effect of treatment observed with the application both at the tillering and anthesis stages, especially under drought stress [63]. Also, field experiment conducted in two seasons in Egypt to evaluate the effect of two nitrogen source combined with foliar spray of Si (diatomite) indicated that organic nitrogen (farm yard manure) combined to diatomite at rate of 0.4% produced the highest values of grain yield, weight of 100 grains and straw yield [64]. In Brazil, foliar application of Si (0.8% of soluble Si, as stabilized orthosilicic acid) increased mass of wheat seed without effect on its germination or vigor [65].

The fertilization results with Si sources on the yield and quality of wheat indicate that there is a trend of significant gains, especially under some kind of stress. However, the results among different studies are variable due to the differences in Si sources, the genetic variations of the wheat cultivars used and the stress levels imposed on the plant.

4. Wheat diseases affected by silicon

The positive effect of Si fertilization on the control of diseases has been reported for pathosystems, mainly involving fungi as pathogens, around the world (**Figure 1**).

For blast (*Magnaporthe oryzae* pathotype Triticum), greenhouse experiments showed reduction of leaf blast severity up to 70% and up to 78% on the area under diseases progress curve (AUDPC) on plants grown in media containing 2 mM of Si compared to plants grown in media without addition of Si source [66, 67]. This effect of Si was associated to the increase in the incubation period by 28% and reduction up to 45% for the number of lesions per cm^2 of leaf [37]. The reduction on blast severity by Si was also associated to the restriction on the host cell colonization by the pathogen [68]. According to authors, in Si-supplied plants the fungal hyphae was restrict to the first-invaded epidermal cell compared to plants not amended with Si in which the fungal hyphae grew successfully and formed an extensive branched mycelium in the first-invaded epidermal cell and several neighboring cells. Leaf application of potassium silicate reduced blast severity, but the positive effect was variable among cultivars [51]. Another study evaluating leaf application of potassium silicate indicated blast severity reduction on the same proportion of fungicide treatment, but no additive or synergistic effect was observed mixing fungicide and potassium silicate [69]. Two-years field experiment showed that Si, applied in the soil as calcium and magnesium silicate in the furrow, and as potassium silicate applied on the leaves, reduced the incidence and severity of blast in the spike, but it effect was variable both with years and cultivars [70].

Powdery mildew (*Blumeria graminis* f. sp. *tritici*) was the first wheat disease reported to be affected by Si [71]. In a three-years experiment evaluating the straw incorporated in the soil conferred the reduction of several wheat disease including

Disease	Si Supply	Environ. (Country)	Variable/Reduction	Ref.
Blast *M. oryzae* Pathotype Triticum	Root	Field (Brazil)	Inc./ 0 to 40% Sev./ 30 to 40%	70
	Foliar	Field (Brazil)	Inc./ 0 to 40% Sev./ 30 to 40%	70
	Root	Greenhouse (Brazil)	AUDPC/ ~31 to 78% Sev./ ~67 to 70%	37, 66, 67, 98
	Foliar	Greenhouse (Brazil)	AUDPC/ ~56	69
Powdery mildew *B. graminis* f.sp. *tritici*	Root	Greenhouse (Brazil, Canada, UK)	Sev./ ~45 to 90%	62, 73-75
		Field (USA)	Sev./ ~29 to 44%	57
	Foliar	Greenhouse (Canada)	Sev./ ~40%	62
Tan spot *P. tritici-repentis*	Root	Greenhouse (Brazil)	AUDPC/ ~73 to 88% Sev./ ~34 to 88%	82 80, 82
		Field (Brazil)	Sev./ ~20 to 50%	unpublished
FBH *F. graminearum*	Root	Greenhouse (Brazil)	Sev./ 20 to 32% AUDPC/ 12 to 21%	83
		Field (Brazil)	Inc./ 0 to 25% Sev./ 0 to 40%	83

Figure 1.
Examples of the effect of silicon (Si) on wheat diseases through root or foliar application (Si supply) in experiments conducted under greenhouse or field environments (Environ.) in different regions (Country) through evaluating disease incidence (Inc.), disease severity (Sev.) or area under disease progress curve (AUDPC) and the percentage of control obtained by silicon treatment. Credits of blast photos to Amanda Baseggio.

powdery mildew [72]. Authors attributed this effect to the increase in the Si availability in the soil conferred by the straw incorporated, being this effect of Si increased in the soil confirmed under greenhouse experiment [72, 73]. Bélanger et al. [74] reported that on wheat plants not supplied with Si the first signs of *B. graminis* f. sp. *tritici* infection were observed at five days after inoculation developing rapidly thereafter reaching to disease severity of up to 40% after five weeks, while plants supplied with Si, colonies of *B. graminis* f. sp. *tritici* were reduced even after five weeks with severity lesser than 5%, indicating very limited fungal colonization on leaf tissue. Later, another study reported reduction on powdery mildew severity up to 80% when Si was supplied via the roots, but leaf spray was less effective reducing the disease severity up to 40% [62]. Field experiment during three consecutive years indicated that calcium silicate (steel slag by-product) reduced powdery mildew severity, in all season that disease occurred, up to 44% [57]. Wheat plants grown in nutrient solution containing different doses of soluble sodium metasilicate showed that the increase of Si concentration in plant showed inverse proportionality to pathogen index indicating an effective action of Si against *B. graminis* f. sp. *tritici* infection in the foliar surface [75].

For spot blotch (*Bipolaris sorokiniana*) the AUDPC was reduced by 59% due soil fertilization with calcium silicate (wollastonite) [76]. The effect of Si on the AUDPC of spot blotch was associated to an increase in the incubation period and decrease in the number of lesions per cm^2 of leaf area and disease severity [77, 78]. This effect of Si on the infectious process of *B. sorokiniana* indicated a limited fungal growth in tissue of Si-supplied plants because authors detected only a sparse network of hypha colonizing the cells as well as a reduced number of epidermal cells showing browning [79].

For tan spot (*Pyrenophora tritici-repentis*), greenhouse experiment using calcium and magnesium silicate (steel slag by-product) incorporated in the soil increased leaf Si concentration which was correlated to longer incubation period and reduced infection efficiency, final number of lesions per cm^2, rate of lesion expansion, lesion size, disease severity and AUDPC [80–82].

For fusarium head blight (*Fusarium graminearum* specie complex), greenhouse experiment indicated that calcium and magnesium silicate incorporated in the soil increased the incubation period in 15% and reduced up to 32% de disease severity and up to 53% de concentration of deoxynivaleonol (a harmful mycotoxin produced by *Fusarium* species) [83]. As the chemical control of fusarium head blight is closely linked to the timing of fungicide application at spike and not all tillers start anthesis at the same time, Si showed a potential to increase the time of fungicide application and still providing a good control of the disease due a longer incubation period and lower rate of colonization (Pazdiora, P. C. Unpublished data).

Field experiment during three years indicated that calcium and magnesium silicate fertilization increased the Si concentration in the soil and wheat leaf and spike tissues, which was associated to the reduction in the severity of both tan spot and fusarium head blight. The reduction of disease severity conferred by Si was greater for tan spot than to fusarium head blight. The greatest control of tan spot and fusarium head blight was obtained with the moderately resistant cultivar treated with two fungicide sprayings. On the other hand, wheat plants grown in soil fertilized with calcium and magnesium silicate that received one application of fungicide at the stem elongation stage showed a reduction up to 50% on tan spot severity and an increase of grain yield by 1 t ha^{-1} compared to the same fungicide treatment on plants grown on soil that received limestone (Pazdiora, P. C. – unpublished data).

Another wheat disease that was affected by Si are leaf blotch (*Parastagnospora nodorum*) under both field and greenhouse trials [72, 73], septoria leaf blotch (*Zymoseptoria tritici*) and eyespot (*Oculimacula yallundae*). However, the efficiency of Si in reducing these diseases was variable and attributed to the type of growing substrate used in the experiments [73]. Furthermore, for bacterial leaf streak (*Xanthomonas translucens* pv. *undulosa*), the Si treatment in the soil not affected the incubation period, latent period, necrotic leaf area, and severity, but reduced up to 50% the chlorotic leaf area [84].

5. Defense responses of wheat activated against pathogens in the presence of Silicon

Several researches have demonstrated the potential of Si in increasing the resistance of wheat against a range of pathogens. Several diseases were reduced on wheat plants supplied with Si through roots or foliar and the mechanism of defense studied. The role of Si on wheat-pathogen interactions is related to its action to increase the plant's defense against to the stressor agent [85].

For blast, in which the pathogen infection leads to increase in the production of reactive oxygen species (ROS) and damage to cell membranes [86], in Si-supplied plants occurred lower concentrations of hydrogen peroxide(H_2O_2) and malondialdehyde indicating, therefore, that the ROS generation and cellular damage were greatly limited [87]. According to authors, the activities of enzymes superoxide dismutase (SOD), catalase (CAT), peroxidase (POX), ascorbate peroxidase (APX) and glutathione-S-transferase (GST) were higher in the leaves of the plants not supplied with Si, while in leaves from Si-supplied plants the glutathione metabolism seemed to play a role in such defense because glutathione reductase activity was increased. In line with this hypothesis, the higher expression levels of the defense-related genes pathogenesis-related 1, chitinase (CHI), POX and phenylalanine ammonia-lyase (PAL), as well as the higher activities of CHI and POX at intermediate and advanced stages of *M. oryzae* infection, respectively, associated to an increase on the concentration of ligninthioglycolic acid derivatives was reported contributing to defense against blast in Si-supplied plants [37, 88]. Cytological and histochemical analysis revealed that in Si-supplied plants the pathogen hyphae were restricted to in the invaded cells, delaying the colonization of the neighboring cells and consequently reducing the progress of the disease [68]. In another study, *M. oryzae* colonization was constrained in the cells on the leaves of Si-supplied plants in association with intense deposition of phenolic-like compounds (flavonoids) [36]. Phenolic-like material was also detected in the parenchyma cells of spikes, and scanning electron micrographs showed that fungal hyphae were scarcely observed in the epidermis, parenchyma and collenchyma cells indicating that these tissues were less colonized by fungal in comparison to the plants not supplied with Si [88].

In the wheat–*B. graminis* f. sp. *tritici* pathosystem, Si increased the resistance to fungus infection by specific defense reactions including papilla formation, production of callose, fungitoxic phenolic compounds and Si deposition at the site of infection [74]. The release of glycosilated phenolics along the cell wall and in association with the compromised haustoria was associated to the degradation of *B. graminis* haustoria [74, 89]. These defense responses potentiated by Si resulted in growth restriction to 10% of epidermal cells and poorly development of haustoria contrasting to leaves of wheat plants not treated with Si that had abundant hyphae of *B. graminis* on epidermal surface and typical haustoria formation in 90% of epidermal cells [74]. The study performed by Rémus-Borel et al. [89] verified that necrotic zones were not detected on *B. graminis* infected tissue, indicating that the response to infection potentiated by Si was not associated with a hypersensitive response, but the newly produced compounds of phenolic-like material that were associated with degraded *B. graminis* haustoria and collapsed conidial chains which interfered with pathogen development. In addition, biochemical defense response *to B. graminis* infection was reported to be associated to production of phytoalexins linked to metabolism of aconitate, which limited the diseases development [90]. A transcriptomic analysis revealed that wheat plants reacted to inoculation with *B. graminis* by an upregulation of many genes linked to stress and metabolic processes and a down-regulation of genes linked to photosynthesis, but in Si-supplied plants the disease development is reduced fact that is translated into a nearly perfect reversal of genes regulated by the effect of *B. graminis* [91]. Another study revealed that *B. graminis* development established a close relationship with the antioxidant response of wheat plants [75]. According authors, the activity of SOD, CAT and APX decreases as Si doses increases indicating a relationship between the applied doses of Si and decrease in *B. graminis* infection due to the reduction of basal antioxidant enzyme activity and ROS. Thus, the decrease of antioxidant enzymes influenced by Si could generate ROS status for more efficient responses of defense to *B. graminis* [75].

On the spot blotch, Si-supply to wheat plants caused a reduction in the rate of infection of *B. sorokiniana* in wheat epidermal cells, due to the physical barrier formed by the cuticle-Si double layer [79]. According to authors, this physical barrier may have reduced the diffusion of lytic enzymes and selective non-host toxins released by the pathogen on the leaf surface, as shown by the reduction of the wax layer degradation. However, even evident the potential of Si accumulated in the plant tissue (cell wall, beneath the cuticle and cell cumulating Si) in inhibiting or delaying the pathogen infection process, this deposition is not homogeneous in the epidermal tissue, which allows the formation of successful infection sites. At this infection sites, others defense potentiated by Si played an important role. Indeed, the increase in the activity of the enzyme POX and increase on the concentration of ligninthioglycolic acid derivatives were related as defense mechanisms, triggered by Si, in the wheat - *B. sorokiana* pathosystem [76].

For tan spot, Si-supply to wheat plants increased biochemical defense mechanisms and histo-cytological defense responses [80, 81]. The most prominent responses from Si-supplied plants were: the accumulation of H_2O_2 in the epidermal cells that occurred early, more intensely and in more epidermal cells, mainly at the beginning of pathogenesis; the alteration of enzyme activities such as SOD, CAT, POX, CHI and PAL; and the accumulation of phenylpropanoid derivatives at the infection site [80, 81]. Together, these defense responses restricted the spread of the pathogen and the damage caused in the plant tissues resulting in a reduction in cell death at *P. tritici-repentis* infection sites [80]. In regarding to the fast and greater accumulation of H_2O_2 in the epidermal cells of the Si-supplied plants is important highlight that the accumulation of H_2O_2 is known to be a mechanism of pathogen attack inducing cell death through *P. tritici-repentis* toxins [92]. However, the early (<12 hours after inoculation) accumulation of H_2O_2 in the epidermal cell of the Si-supplied plants of moderately resistant cultivar, compared to late accumulation (>24 after inoculation) in the mesophyll and epidermal cells of the non supplied plants, indicated that H_2O_2 was a defense mechanism. This inference is because accumulation of H_2O_2 occurred before pathogen penetration into the leaf tissue and was related to lower infection efficiency (the ratio between the number of conidia on the leaf surface and the number of lesions formed). Furthermore, on the Si-supplied plants, early fluorescence in epidermal cells, in neighboring cells and in the cell in which *P. tritici-repentis* attempted to penetrate, indicated that phenylpropanoid derivative accumulation were also contributing to disease resistance [81].

6. Physiological effects of silicon in wheat under pathogen stress

The photosynthesis is the major physiological process in plants; therefore, if plants are infected by pathogens some process in their physiology can be negatively affected. The pathogen infection can be responsible to decrease photosynthesis at different levels [93], modification or damage of the photosynthetic apparatus [94] and interfering with normal source-sink relationships in plants [95, 96].

In this sense exist a general consensus that Si improves the plant resistance to various biotic and abiotic stresses. Thus, the effect of Si on plant physiology it has been observed mainly when plant is under some kind of stress. For example, under biotic stress imposed by *B. graminis* f. sp. *tritici*, an analysis of around 55,000 transcripts indicated that around 3000 genes were differentially expressed on pathogen-inoculated plants, but a nearly perfect reversal in the transcript profile of downregulated stress-related genes occurred when Si was supplied [91]. This result indicated that Si rather than being involved directly in the regulation of gene expression, prevented or attenuated the effects on transcription imposed by pathogen [91]. Furthermore, several studies revealed that wheat plants supplied with Si when challenged by pathogens showed lower affectation and/or ameliorative on photosynthetic process as assessed via measurements of the leaf gas exchange and the chlorophyll *a* (Chl*a*) fluorescence kinetics.

In this regard, some studies showed that concentration of photosynthetic pigments and structural and functional damage of chloroplasts produce alterations on photochemical machinery with losses in the amount of chlorophylls and carotenoids, as a result it has been observed decreased values for the net photosynthesis rate [93]. In a study on wheat-*Magnaporthe oryzae* interaction, in Si-supplied plants occurred a maintaining the concentration of photosynthetic pigments such as total chlorophyll, violanxanthin + antheraxanthin + zeaxanthin, β-carotene and α-carotene which helped to maintain the structural and functional viability of the photosynthetic machinery minimizing, therefore, lipid peroxidation and the production of ROS to ensure the integrity of the leaf cells [97]. In the same pathosystem, photosynthetic performance was studied in Si-supplied plants which showed higher values for net photosynthesis rate coupled with improved photochemistry associated to Chl*a* fluorescence parameters, and also increased concentrations of total chlorophylls [66, 98]. Also, Si-supplied plant showed less functional damage to the photosystem II (PSII) without reductions in the values of maximum quantum quenching, photochemical yield of PSII and electron transport rate, but higher values for quenching non-photochemical [97].

Likewise, the impairment caused by blast on the photosynthetic process, primarily related to the F_v/F_m parameter, on wheat leaves, was in lesser extent on the plants sprayed with potassium silicate [69]. Furthermore, authors did not detect any significant alteration on the gas exchange and Chl*a* fluorescence parameter for plants sprayed three times (every 96 h interval) as the potassium silicate rates increased from 2.5 to 12.5 g L^{-1} indicating that potassium silicate do not cause perturbation to the photosynthetic machinery of wheat plants.

In addition, the pathogen infection usually leads to the development of symptoms that result in a decrease on the photoassimilates production [99], resulting in low performance of photochemical reactions associated to PSII, that mainly influence the reduction in CO_2 assimilation [93] producing alteration in some parameters of leaf gas exchange. In this way, alterations with diffusional limitations and significant losses both in the electron transport rate and biochemical capacity for carboxylation associated with losses in ribulose-1,5-bisphosphate carboxylase/oxygenase (Rubisco) activity have been measured on the leaves of wheat plants infected with *M. oryzae* [87]. In this pathosystem, Aucique Perez [66] observed that these dysfunctions could largely be avoided in the presence of Si, which might directly be associated with lower blast symptoms on Si-supplied plants; in those plants net CO_2 assimilation rate, stomatal conductance to water vapor, and transpiration rate were significantly higher, showing that Si improving gas exchange performance. Furthermore, Araujo et al. [95] measured several parameters of Chl*a* fluorescence, sugars (glucose, fructose and sucrose) and starch concentration, the activities of enzymes acid invertase and sucrose phosphate synthase in leaves and spike of wheat challenged by *M. oryzae* showing evidences of the beneficial effects of Si in improving the source-sink relationship on infected leaves and spikes by preserving the alteration in assimilates production and partitioning during the grains filling process.

Overall, in all the experiments the authors agreed that the effect of Si on photosynthesis process is major in plants challenged by the pathogen. Indeed, transcriptomic studies performed on several plant species submitted to different types of biotic stress showed a reduction in transcript levels of genes related to photochemistry, Calvin cycle and the synthesis of chlorophylls [100]. Observations in non-inoculated plants, in general, does no showed significant difference between the non-supplied and Si-supplied plants for the values of leaf gas exchanges, photochemical parameters associated with Chl*a* fluorescence, soluble sugars and some enzymes of sucrose metabolism [95]. These findings are in line with previous study in which was not found any differences on the photosynthetic activity of rice plants with and without Si supply [46]. In this context, Coskun et al. [14] pointing out to the question of Si's role in the absence of stress having little or no effect, however remains a contentious issue. Probably Si is indirectly involved in the nutrition of the plant and it is undeniable that Si prevents or mitigates the strains imposed by stress, and this, ultimately, is reflected in improvements in plant growth, function and metabolic activity.

7. Conclusion and perspectives

The importance of Si and Si fertilization for improving plant health are recognized. Several studies clearly demonstrate that silicate fertilization for wheat plants increases grain yield and its quality, especially under both biotic and abiotic stress. In the case of biotic stress, the Si effect reduces the intensity of the diseases due to the enhancement of the defense mechanisms that are earlier expressed and better coordinated. In addition to the effect on defenses against the pathogen, plants supplied with Si also show less physiological damage, in fact this is associated with increased on the yield. These effects are clearly evident and largely accepted, indicating that Si fertilization could be incorporated in the wheat management.

Nevertheless, for silicate fertilization to become widely used by wheat growers, several issues still need to be clarified. Initially, an important point is that most of the studies demonstrating the effect of Si on disease control and the reduction of physiological damage was carried out in a controlled environment with only a single stress imposed on the plant, and few studies were carried out in field conditions with multiple stresses simultaneously. Therefore, more studies need to be carried out under field conditions to obtain a greater amount of data of the silicate fertilization effect and with all these data should be analyzed through meta-analyzes to provide a holistic view of the effect.

Taking into account that we can mostly use two forms of Si application: leaf or root, there are still many unanswered questions. For example, in soil fertilization we can consider the following questions.

1. What is the best form of application? In this sense, we can consider situations in which the farmer plows the soil and the silicate fertilizer can be incorporated during this procedure. However, for wheat growers who use no-till, incorporation is not possible. In this case, the application of the Si source can be carried out on the soil surface or in the sowing line. For application to the surface without incorporation, we still do not know clearly how long it takes for Si to be available in the soil solution in sufficient quantity to meet the demand of the wheat plant. With respect to the application in the sowing line, there is still not enough data to indicate which is the best source or dose of Si to supply to the plant without interfering in the initial stages of seedling development, and also is still unknown which is the amount to applied without compromising the logistic yield of the sowing procedure to obtain the Si benefits as well as possible.

2. How often should silicate fertilization be carried out? Evidently, this information can be obtained by analyzing the amount of Si available in the soil. However, for the wheat growers to adopt silicate fertilization it will be necessary to know the cost benefit of the application and the frequency of application. In the case of fertilization in the sowing line, the financial impact for the producer is easier to be determined. However, in surface or incorporated applications, where specific activities are required for this procedure, more information is needed. For example, how many crop cycles/years should the reapplication be carried out? What dose should be applied and/or reapplied? What is the best product for reapplication: soluble or powder? Can we make a basic application to increase the Si pool in the soil and the reapplications be carried out via the seeding line? Does crop rotation or succession affect the frequency and/or rate that we should be used when reapplying silicate fertilization? These are questions that remain unanswered to wheat growers.

3. Considering the great variation in the ability of Si absorption among different wheat cultivars, it is important that this factor to be considered in breeding programs, aiming to obtain cultivars that present a higher efficiency Si absorption for different soils and climates it will be expected. This is important to maximize the use of silicate fertilization and consequently maximize the economic return to the producer. Furthermore, according to Ranjbar et al. [30] the selection and modification of silicon-efficient wheat cultivars can be a successful and promising strategy to maintain production in low-input and environmentally friendly agricultural systems.

4. With regard to foliar application, we agree with the consideration pointed out by Puppe and Sommer [101] that, there is little knowledge on Si foliar application and Si fertilizers for different purposes (biotic and abiotic stress). The foliar application needs further detailed studies, especially on the knowledge on concentrations of foliar Si fertilizers application, type of fertilizers, frequency of application and the timing of spraying.

Further research should be done to answer these questions, even though we will be closer to being able to clearly demonstrate to wheat growers the real benefit, in economic terms, and the routinely adopt silicate fertilization for wheat crop.

Acknowledgements

The authors are thankful to the CAPES (Finance code 001). L.J. Dallagnol is supported by fellowship of the Brazilian National Council for Scientific and Technological Development (CNPq) (grant number 308149/2018-1). This work has been partially supported by the Brazilian agency CAPES.

Conflict of interest

The authors declare no conflict of interest.

Author details

Leandro José Dallagnol[1*], Andrea Elizabeth Román Ramos[1,2] and Keilor da Rosa Dorneles[1]

1 Crop Protection Department, Agronomy Faculty EliseuMaciel, Federal University of Pelotas, Pelotas, Brazil

2 Agricultural Sciences Natural Resources and the Environment Faculty, Bolivar State University, Guaranda, Ecuador

*Address all correspondence to: leandro.dallagnol@ufpel.edu.br

References

[1] FAO- Food and Agriculture Organization. FAOSTAT - Crops. [Internet]. 2020. Available from: http://www.fao.org/faostat/en/#data/QC/ [Accessed: 2020-19-13]

[2] Tadesse W, Halila H, Jamal M, El-Hanafi S, Assefa S, Oweis T, Baum M. Mint: Role of sustainable wheat production to ensure food security in the CWANA region. Journal of Experimental Biology and Agricultural Sciences. 2017;5:S15-S32. DOI: 10.18006/2017.5(Spl-1-SAFSAW).S15.S32

[3] Chaves MS, Martinelli JÁ, Wesp-Guterres C, Graichen FAZ, Brammer SP, Scagliusi SM, Consoli L. The importance for food security of maintaining rust resistance in wheat. Food Security. 2013;5:157-176. DOI: 10.1007/s12571-013-0248-x

[4] Shiferaw B, Smale M, Braun H-J, Duveiller E, Reynolds M, Muricho G. Crops that feed the world 10. Past successes and future challenges to the role played by wheat in global food security. Food Security. 2013;5:291-317. DOI: 10.1007/s12571-013-0263-y

[5] Acevedo M, Zurn JD, Molero G, Singh P, He X, Aoun M, Juliana P, Bochleman H, Bonman M, El-Sohl M, Amri A, Coffman R, McCandless L. The role of wheat in global food security. In: Nagothu US, editor. Agricultural Development and Sustainable Intensification: Technology and Policy Challenges in the Face of Climate Change. 1st ed. New York: Routledge; 2018. p. 81-110. DOI: 10.4324/9780203733301-4

[6] Bai G, Su Z, Cai J. Wheat resistance to Fusarium head blight. Canadian Journal of Plant Pathology. 2018;40:336-346. DOI: 10.1080/07060661.2018.1476411

[7] Faris JD, Liu Z, Xu SS. Genetics of tan spot resistance in wheat. Theoretical and Applied Genetics. 2013;126:2197-2217. DOI: 10.1007/s00122-013-2157-y

[8] Goddard R, Steed A, Chinoy C, Ferreira JR, Scheeren PL, Maciel JLN, Caierão E, Torres GAM, Consoli L, Santana FM Fernandes JMC, Simmonds J, Uauy C, Cockram J, Nicholson P. Dissecting the genetic basis of wheat blast resistance in the Brazilian wheat cultivar BR 18-Terena. BMC Plant Biology. 2020;20:1-15. DOI: 10.1186/s12870-020-02592-0

[9] Periyannan S, Milne RJ, Figueroa M, Lagudah ES, Dodds PN. An overview of genetic rust resistance: From broad to specific mechanisms. PLoS Pathogens. 2017;13: e1006380. DOI: 10.1371/journal.ppat.1006380

[10] Santana FM, Lau D, Sbalcheiro CC, Goussain RCS, Venancio WS, Custódio AAP, Moreira LSO, Sussel AAB, editors. Eficiência de fungicidas para controle de brusone de trigo: resultados dos ensaios cooperativos, safra 2018.53nd ed. Embrapa Trigo. Circular Técnica Online. Passo Fundo, RS: 2020. 18 p.

[11] Santana FM, Lau D, Sbalcheiro CC, Schipanski CA, Venancio WS, Dallagnol LJ, Guterres CW, Kuhnem Junior PR, Chagas DF, editors. Eficiência de fungicidas para controle de giberela do trigo: resultados dos Ensaios Cooperativos - Safra 2018. 52nd ed. Embrapa Trigo. Circular Técnica Online. Passo Fundo, RS: 2020. 20 p.

[12] Frac-Fungicide Resistance Action Committee. Pathogen risk list [Internet]. 2019. Available from: https://www.frac.info/docs/default-source/publications/pathogen-risk/frac-pathogen-list-2019.pdf. [Accessed: 2020-19-13]

[13] Haynes RJ. A contemporary overview of silicon availability in agricultural soils. Journal of Plant Nutrition and Soil Science. 2014;177:831-844. DOI: 10.1002/jpln.201400202

[14] Coskun D, Deshmukh R, Sonah H, Menzies JG, Reynolds O, Ma JF, Kronzucker HJ, Bélanger RR. The controversies of silicon's role in plant biology. New Phytologist. 2019;221:67-85. DOI: 10.1111/nph.15343

[15] Debona D, Rodrigues FA, Datnoff LE. Silicon's role in abiotic and biotic plant stresses. Annual Review of Phytopathology 2017;55:85-107. DOI: 10.1146/annurev-phyto-080516-035312

[16] Luyckx M, Hausman JF, Lutts S, Guerriero G. Impact of silicon in plant biomass production: Focus on bast fibres, hypotheses, and perspectives. Plants. 2017;6:37. DOI: 10.3390/plants6030037

[17] Rodrigues FA, Dallagnol LJ, Datnoff LE. Silicon control of foliar diseases in monocots and dicots. In Rodrigues FA, Datnoff LE, editors. Silicon and Plant Diseases. 1 ed. Berlin: Springer; 2015. p. 67-108. https://doi.org/10.1007/978-3-319-22930-0_4

[18] Chiba Y, Mitani N, Yamaji N, Ma JF. HvLsi1 is a silicon influx transporter in barley. The Plant Journal. 2009;57:810-818. DOI:10.1111/j.1365-313X.2008.03728.x

[19] Ma JF, Yamaji N. A cooperative system of silicon transport in plants. Trends in Plant Science. 2015; 20:435-442. DOI: 10.1016/j.tplants.2015.04.007.

[20] Mitani N, Yamaji N, Ago Y, Iwasaki K, Ma JF. Isolation and functional characterization of an influx silicon transporter in two pumpkin cultivars contrasting in silicon accumulation. Plant Journal. 2011;66:231-40. DOI: 10.1111/j.1365-313X.2011.04483.x

[21] Montpetit J, Vivancos J, Mitani-Ueno N, Yamaji N, Rémus-Borel W, Belzile F, Ma JF, Bélanger RR. Cloning, functional characterization and heterologous expression of TaLsi1, a wheat silicon transporter gene. Plant Molecular Biology. 2012;79:35-46. DOI: 10.1007/s11103-012-9892-3

[22] Jarvis SC. The uptake and transport of silicon by perennial ryegrass and wheat. Plant and Soil 1987;97:429-437. DOI: 10.1007/BF02383233

[23] Sangster AG, Hodson MJ. Silica deposition in subterranean organs. In: Rapp Jr G, Mulholland SC, editors. Phytolith Systematics: Emerging Issues. Advances in Archaeological and Museum Science. 1 ed. Springer US; 1992. p. 239-251. DOI: 10.1007/978-1-4899-1155-1

[24] Mayland HF, Wright JL, Sojka, RE. Silicon accumulation and water uptake by wheat. Plant and Soil 1991; 137:191-199. DOI: 1007/BF00011197

[25] Rafi MM, Epstein E. Silicon absorption by wheat (*Triticum aestivum* L). Plant and Soil1999;211:223-230. DOI: 10.1023/A:1004600611582

[26] Rains DW, Epstein E, Zasoski RJ, Aslam M. Active Silicon Uptake by Wheat. Plant and Soil 2006; 280: 223-228. DOI:1007/s11104-005-3082-x

[27] Gomes D, Agasse A, Thiebaud P, Delrot S, Geros H, Chaumont F. Aquaporins are multifunctional water and solute transporters highly divergent in living organisms. Biochimica et Biophysica Acta (BBA)-Biomembranes. 2009;1788:1213-1228. DOI: 10.1016/j.bbamem.2009.03.009

[28] Murozuka E,Bang TC, Frydenvang J, Lindedam J, Laursen KH, Bruun S, Magid J, Schjoerring JK. Concentration of mineral elements in wheat (*Triticum aestivum* L.) straw: genotypic differences and consequences for enzymatic saccharification. Biomass and Bioenergy 2015;75:134-141. DOI: 10.1016/j.biombioe.2015.02.017

[29] Carter AH, Rath BB, Gorzkowski EP, Qadri SB. Evaluation of silica content in winter wheat chaff. Agricultural &

Environmental Letters. 2020;5:e20025. DOI:10.1002/ael2.20025

[30] Ranjbar SS, Motesharezadeh B, Moshiri F, Mirseyed Hosseini H, Alikhani, HA. Silicon utilization efficiency of different wheat cultivars in a calcareous soil. SILICON 2019;11:2159-2168. DOI: 10.1007/s12633-018-0038-3

[31] Mandlik R, Thakral V, Rature G, Shinde S, Nikolic M, Tripathi DK, Sonah H, Deshmukh R. Significance of silicon uptake, transport, and deposition in plants. Journal of Experimental Botany. 2020, DOI:10.1093/jxb/eraa301

[32] Raven JA. Cycling silicon–the role of accumulation in plants. New Phytologist. 2003;158:419-421. DOI: 10.1046/j.1469-8137.2003.00778.x

[33] Tripathi, D.K., Kumar, R., Pathak, A.K, Chauhan DK, Rai AK. Laser-Induced Breakdown Spectroscopy and Phytolith Analysis: An Approach to Study the Deposition and Distribution Pattern of Silicon in Different Parts of Wheat (*Triticum aestivum* L.) Plants & Agriculture Research. 2012;1:352-361. DOI: 10.1007/s40003-012-0042-6

[34] Neu S, Schaller J, Dudel EG. Silicon availability modifies nutrient use efficiency and content, C:N:P stoichiometry, and productivity of winter wheat (*Triticum aestivum* L.). Scientific Reports 2017;7:40829. DOI:10.1038/srep40829

[35] Peleg Z, Saranga Y, Fahima T, Aharoni A, Elbaum R. Genetic control over silica deposition in wheat awns. Physiologia Plantarum 2010;140:10-20. DOI:10.1111/j.1399-3054.2010.01376.x

[36] Silva WL, Cruz MFA, Fortunato AA, Rodrigues FÁ. Histochemical aspects of wheat resistance to leaf blast mediated by silicon. Scientia Agricola 2015;72:322-327. DOI:10.1590/0103-9016-2014-0221

[37] Xavier Filha MS, Rodrigues FA, Domiciano GP, Oliveira HV, Silveira PR, Moreira WR. Wheat resistance to leaf blast mediated by silicon. Australasian Plant Pathology. 2011;40:28-38. DOI:10.1007/s13313-010-0010-1

[38] Mecfel J, Hinke S, Goedel WA, Marx G, Fehlhaber R, Bäucker E, Wienhaus O. Effect of silicon fertilizers on silicon accumulation in wheat. Journal of Plant Nutrition and Soil Science. 2007;170:769-772. DOI:10.1002/jpln.200625038

[39] Guntzer F, Keller C, Poulton PR, McGrath SP, Meunier JD. Long-term removal of wheat straw decreases soil amorphous silica at Broadbalk, Rothamsted. Plant and Soil 2012;352:173-184. DOI: 10.1007/s11104-011-0987-4

[40] Tubana BS, Babu T, Datnoff LE. A Review of silicon in soils and plants and its role in US agriculture: History and future perspectives. Soil Science. 2016;181:393-411. DOI: 10.1097/SS.0000000000000179

[41] Epstein S. Integration of the cognitive and the psychodynamic unconscious. American Psychologist. 1994;49:709-724. DOI: 10.1037//0003-066X.49.8.709

[42] Maxim LD, Niebo R, Larosa S, Johnston B, Allison K, McConnell EE. Product stewardship inwollastonite production. Inhalation Toxicology. 2008; 20:1199-1214. DOI: 10.1080/08958370802136749

[43] Buck GB, Korndorfer GH, Datnoff LE. Extractors for estimating plant available silicon from potential silicon fertilizer sources. Journal of Plant Nutrition. 2010; 34:272-282. DOI: 10.1080/01904167.2011.533327

[44] Sebastian D, Rodrigues H, Kinsey C, Korndörfer G, Pereira H, Buck G, Datnoff L, Miranda S, Provance-Bowley M. A 5-day method for determination of soluble silicon concentrations in nonliquid fertilizer

materials using a sodium carbonate-ammonium nitrate extractant followed by visible spectroscopy with heteropoly blue analysis: Single-laboratory validation. Journal of AOAC INTERNATIONAL. 2013;96:251-258. DOI: 10.5740/jaoacint.12-243

[45] Wang M, Jim JW, Xudong W. Effect of KOH-enhanced biochar on increasing soil plant-available silicon. Geoderma 2018; 321: 22-31. DOI: 10.1016/j.geoderma.2018.02.001

[46] Ma JF, Takahashi E, editors. Soil, fertilizer, and plant silicon research in Japan. 1 ed. Elsevier: Amsterdam, Netherlands; 2002. 294 p. DOI: 10.1016/B978-0-444-51166-9.X5000-3

[47] Nolla A, Korndorfer GH, Da Silva CAT, Da Silva TRB, Zucarelli V, Da Silva MAG. Correcting soil acidity with the use of slags. African Journal of Agricultural Research. 2013;8:5174-5180. DOI: 10.5897/AJAR2013.6940

[48] Alvarez J, Snyder GH, Anderson DL, Jones DB. Economics of calcium silicate slag in a rice-sugarcane rotation in the everglades. Agricultural Systems 1988; 28:179-188. DOI: 10.1016/0308-521X(88)90050-9

[49] Ning DF, Song A, Fan FL, Li ZJ, Liang YC. Effects of slag-based silicon fertilizer on rice growth and brown-spot resistance. PLoSOne. 2014;9:e102681. DOI: 10.1371/journal.pone.0102681

[50] Raid RN, Anderson DL, Ulloa MF. Influence of cultivar and amendment of soil with calcium silicate slag on foliar disease development and yield of sugar-cane. Crop Protection. 1992;11: 84-88. DOI: 10.1016/0261-2194(92)90085-J

[51] Cruz MFA, Diniz APC, Rodrigues FA, Barros EG. Aplicação foliar de produtos na redução da severidade da brusone do trigo. Tropical Plant Pathology. 2011;36:424-428. DOI: 10.1590/S1982-56762011000600014

[52] Frayssinet C, Osterrieth LM, Borrelli LN, Honaine MF, Ciarlo E, Heiland P. Effect of silicate fertilizers on wheat and soil properties in Southeastern Buenos Aires province, Argentina. A preliminary study. Soil and Tillage Research. 2019;195:104412. DOI: 10.1016/j.still.2019.104412

[53] Gocke M, Liang W, Sommer M, Kuzyakov Y. Silicon uptake by wheat: effects of Si pools and pH. Journal of Soil Science and Plant Nutrition. 2013;176:551-560. DOI: 10.1002/jpln.201200098

[54] Mousavi SR, Rezaie M. Nanotechnology in agriculture and food production. Journal of Applied Environmental and Biological Sciences. 2011;10:414-419.

[55] Liang YC, Ma TS, Li FJ, Feng YJ. Silicon availability and response of rice and wheat to silicon in calcareous soils. Communications in Soil Science and Plant Analysis. 1994;25:2285-2297. DOI: 10.1080/00103629409369189

[56] Wang HL, Li CH, Liang YC. Agricultural utilization of silicon in China. In: Datnoff LE, Snyder GH, Korndörfer GH, editors. In: Silicon in agriculture. Studies in plant science, 8 ed. Amsterdam: Elsevier; 2001. p. 343-352. DOI: 10.1016/S0928-3420(01)80001-X

[57] Provance-Bowley MC, Heckman JR, Durner EF. Calcium silicate suppresses powdery mildew and increases yield of field grown wheat. Soil Science Society of America Journal. 2010;74:1652-1661. DOI:10.2136/sssaj2010.0134

[58] Walsh OS, Shafian S, McClintick-Chess JR, Belmont KM, Blanscet SM. Potential of silicon amendment for improved wheat production. Plants. 2018;28:26. DOI: 10.3390/plants7020026

[59] White B, Tubana BS, Babu T, Mascagni H, Agostinho F, Datnoff LE, Harrison S. Effect of silicate slag application on wheat grown under two

nitrogen rates. Plants. 2017;11:47. DOI: 10.3390/plants6040047

[60] Sarto MVM, Lana MC, Rampim L, Rosset JSR, Wobeto JR. Effects of silicate application on soil fertility and wheat yield. Semina: Ciências Agrárias. 2015; 36:4071-4082. DOI: 10.5433/1679-0359.2015v36n6Supl2p4071

[61] Kowalska J, Tyburski J, Jakubowska M, Krzymińska J. Effect of different forms of silicon on growth of spring wheat cultivated in organic farming system. SILICON 2020. DOI: 10.1007/s12633-020-00414-4

[62] Guével M, Menzies JG, Bélanger RR. Effect of root and foliar applications of soluble silicon on powdery mildew control and growth of wheat plants. European Journal of Plant Pathology. 2007;119:429-436. DOI: 10.1007/s10658-007-9181-1

[63] Maghsoudi K, Emam Y, Ashraf M. Foliar application of silicon at different growth stages alters growth and yield of selected wheat cultivars. Journal of Plant Nutrition. 2016;39:1194-1203. DOI: 10.1080/01904167.2015.1115876

[64] Hellal FA, Zeweny RM, Yassen AA. Evaluation of nitrogen and silicon application for enhancing yield production and nutrient uptake by wheat in clay soil. Journal of Applied Sciences Research. 2012;8:686-692.

[65] Toledo MZ, Castro GSA, Crusciol CAC, Soratto RP, Cavariani C, Ishizuka MS, Picoli LB. Silicon leaf application and physiological quality of white oat and wheat seeds. Semina: Ciências Agrárias. 2012;33:1693-1702. DOI: 10.5433/1679-0359.2012v33n5p1693

[66] Aucique Perez CE, Rodrigues FA, Moreira WR, DaMatta FM. 2014. Leaf gas exchange and chlorophyll a fluorescence in wheat plants supplied with silicon and infected with *Pyricularia oryzae*. Phytopathology 2014;104:143-149. DOI: 10.1094/PHYTO-06-13-0163-r

[67] Debona D, Rodrigues FA, Rios JA, Nascimento KJT, Silva LC. The effect of silicon on antioxidant metabolism of wheat leaves infected by *Pyricularia oryzae*. Plant Pathology. 2014;63:581-589. DOI: 10.1111/ppa.12119

[68] Sousa RS, Rodrigues FA, Schurt DA, Souza NFA, Cruz MFA. Cytological aspects of the infection process of *Pyricularia oryzae* on leaves of wheat plants supplied with silicon. Tropical Plant Pathology. 2013;38:472-477. DOI: 10.1590/S1982-56762013000600002

[69] Oliveira, TB, Aucique-Pérez, CE, Rodrigues, FÁ. Foliar application of silicon decreases wheat blast symptoms without impairing photosynthesis. Bragantia 2019;78:423-431. DOI: 10.1590/1678-4499.20180379

[70] Pagani APS, Dianese AC, Café-Filho AC. Management of wheat blast with synthetic fungicides, partial resistance and silicate and phosphite minerals. Phytoparasitica 2014;42:609-617. DOI:10.1007/s12600-014-0401-x

[71] Germar B. Über einige Wirkungen der Kieselsäure in Getreidepflanzen, insbesondere auf deren Resistenz gegenüber Mehltau. Z. Pflanzenernaehr. Dueng. Bodenk. 1934;35:102-115. DOI: 10.1002/jpln.19340350111

[72] Rodgers-Gray BS, Shaw MW. Substantial reductions in winter wheat diseases caused by addition of straw but not manure to soil. Plant Pathology. 2000;49: 590-599. DOI:10.1046/j.1365-3059.2000.00497.x

[73] Rodgers-Gray BS, Shaw MW. Effects of straw and silicon soil amendments on some foliar and stem-base diseases in pot-grown winter wheat. Plant Pathology. 2004;53:733-740. DOI:10.1111/j.1365-3059.2004.01102.x

[74] Bélanger RR, Benhamou N, Menzies JG. Cytological evidence of an active role of silicon in wheat resistance to powdery mildew (*Blumeria graminis* f.

sp. tritici). Phytopathology 2003;93:402-412. DOI: 10.1094/PHYTO.2003.93.4.402

[75] Moldes CA, Lima Filho OF, Merini LJ, Tsai SM, Camiña JM. Occurrence of powdery mildew disease in wheat fertilized with increasing silicon doses: a chemometric analysis of antioxidant response. Acta Physiologiae Plantarum. 2016;8:206-238. DOI: 10.1007/s11738-016-2217-4

[76] Domiciano GP, Rodrigues FA, Moreira WR, Oliveira HV, Vale FXR, Xavier Filha MS. Silício no progresso da mancha marrom na folha bandeira do trigo. Tropical Plant Pathology. 2010;35:186-189. DOI:10.1590/S1982-56762010000300009

[77] Domiciano GP, Rodrigues FA, Vale FXR, Xavier Filha MS, Moreira WR, Andrade CCL, Pereira SC. Wheat resistance to spot blotch potentiated by silicon. Journal of Phytopathology 2010;158:334-343. DOI: 10.1111/j.1439-0434.2009.01623.x

[78] Zanão Júnior LA, Fontes RLF, Coelho PHM, Korndörfer GH, Zambolim L. Silício aplicado no solo reduz a severidade da mancha-marrom do trigo em solos com baixos teores desse elemento. Revista Brasileira de Ciência do Solo. 2010;34:401-408. DOI: 10.1590/S0100-06832010000200013

[79] Domiciano GP, Rodrigues FA, Guerra AMN, Vale FXR. Infection process of *Bipolaris sorokiniana* on wheat leaves is affected by silicon. Tropical Plant Pathology.2013;38:258-263. DOI:10.1590/S1982-56762013005000006

[80] Dorneles KR, Dallagnol LJ, Pazdiora PC, Rodrigues FA, Deuner S. Silicon potentiates biochemical defense responses of wheat against tan spot. Physiological and Molecular Plant Pathology. 2017;97:69-78. DOI: 10.1016/j.pmpp.2017.01.001

[81] Dorneles KR, Pazdiora PC, Hoffmann JF, Chaves FC, Monte LG, Rodrigues FA, Dallagnol LJ. Wheat leaf resistance to *Pyrenophora tritici-repentis* induced by silicon activation of phenylpropanoid metabolism. Plant Pathology. 2018;67:1713-1724. DOI: 10.1111/ppa.12876

[82] Pazdiora PC, Dorneles KR, Forcelini CA, Del Ponte EM, Dallagnol LJ. Silicon suppresses tan spot development on wheat infected by *Pyrenophora tritici-repentis*. European Journal of Plant Pathology. 2018;150:49-56. DOI: 10.1007/s10658-017-1251-4

[83] Pazdiora, P.C. Silício, resistência parcial e fungicida no manejo da giberela do trigo (Thesis). Pelotas: Federal University of Pelotas; 2019.

[84] Silva IT, Rodrigues F, Oliveira JR, Pereira SC, Andrade CCL, Silveira PR, Conceição MM. Wheat resistance to bacterial leaf streak mediated by silicon. Journal of Phytopathology 2010; 158: 253-262. DOI: 10.1111/j.1439-0434.2009.01610.x

[85] Islam W, Tayyab M, Khalil F, Hua Z, Huang Z, Chen HTH. Silicon-mediated plant defense against pathogens and insect pests. Pesticide Biochemistry and Physiology. 2020;168:104641. DOI:10.1016/j.pestbp.2020.104641

[86] Debona D, Rodrigues FA, Rios JA, Nascimento KJT. Biochemical changes in the leaves of wheat plants infected by *Pyricularia oryzae*. Phytopathology 2012; 102:1121-1129. DOI: 10.1094/PHYTO-06-12-0125-r

[87] Debona D, Rodrigues FÁ, Rios JA, Martins SC, Pereira LF, DaMatta FM. Limitations to photosynthesis in leaves of wheat plants infected by *Pyricularia oryzae*. Phytopathology 2014;104:34-39. DOI: 10.1094/PHYTO-01-13-0024-r

[88] Cruz MFA, Debona D, Rios JA, Barros EG, Rodrigues FA. Potentiation of defense-related gene expression by silicon increases wheat resistance to leaf blast. Tropical

Plant Pathology.2015;40:394-400. DOI:10.1007/s40858-015-0051-7

[89] Rémus-BoreL W, Menzies JG, Belanger RR. Silicon induces antifungal compounds in powdery mildew-infected wheat. Physiological and Molecular Plant Pathology.2005;66:108-115. DOI: 10.1016/j.pmpp.2005.05.006

[90] Rémus-Borel W, Menzies JG, Belanger RR. Aconitate and methyl aconitate are modulated by silicon in powdery mildew-infected wheat plants. Journal of Plant Physiology 2009;166:1413-1422. DOI: 10.1016/j.jplph.2009.02.011

[91] Chain F, Côté-Beaulieu C, Belzile F, Menzies JG, Bélanger RR. A comprehensive transcriptomic analysis of the effect of silicon on wheat plants under control and pathogen stress conditions. Molecular Plant-Microbe Interactions. 2009; 22:1323-1330. DOI: 10.1094/MPMI-22-11-1323

[92] Ciuffetti LM, Manning VA, Pandelova I, Betts MF, Martinez JP. Host-selective toxins, PtrToxA and PtrToxB, as necrotrophic effectors in the *Pyrenophora tritici-repentis*–wheat interaction. New Phytologist. 2010;187:911-919. DOI:10.1111/j.1469-8137.2010.03362.x

[93] Berger S, Benediktyova Z, Matous K, Bonfig KB, Mueller MJ, Nedbal L, Roitsch T. Visualization of dynamics of plant–pathogen interaction by novel combination of chlorophyll fluorescence imaging and statistical analysis: differential effects of virulent and avirulent strains of *Pseudomonas syringae* and of oxylipins on *Arabidopsis thaliana*. Journal of Experimental Botany. 2007;58:797-806. DOI: 10.1093/jxb/erl208

[94] Lichtenthaler HK, Miehé JA. Fluorescence imaging as a diagnostic tool for plant stress. Trends in Plant Science. 19972:316-320. DOI: 10.1016/S1360-1385(97)89954-2

[95] Araújo MUP, Rios JA, Silva ET, Rodrigues FÁ. Silicon alleviates changes in the source-sink relationship of wheat plants infected by *Pyricularia oryzae*. Phytopathology 2019;109:1129-1140. DOI: 10.1094/PHYTO-11-18-0428-r

[96] Biemelt S, Sonnewald U. Plant-microbe interactions to probe regulation of plant carbon metabolism. Journal of Plant Physiology 2006;163:307-318. DOI: 10.1016/j.jplph.2005.10.011

[97] Aucique-Pérez CE, Menezes Silva PE, Moreira WR, DaMatta FM, Rodrigues FÁ. Photosynthesis impairments and excitation energy dissipation on wheat plants supplied with silicon and infected with *Pyricularia oryzae*. Plant Physiology and Biochemistry. 2017;121: 196-205. DOI: 10.1016/j.plaphy.2017.10.023

[98] Rios J, Rodrigues F, Debona D, Silva L. Photosynthetic gas exchange in leaves of wheat plants supplied with silicon and infected with *Pyricularia oryzae*. Acta Physiologia Plantarum. 2014;36:371-379. DOI: 10.1007/s11738-013-1418-3

[99] Schreiber U.Pulse-Amplitude-Modulation (PAM) fluorometry and saturation pulse method: An overview. In: Papageorgiou GC, Govindjee, editors. Chlorophyll Fluorescence: A Signature of Photosynthesis. 1nd ed. Springer Verlag: Dordrecht; 2004. p. 279-319. DOI: 10.1007/978-1-4020-3218-9

[100] Bilgin DD, Zavala JA, Zhu JI, Clough SJ, Ort DR, DeLucia EH. Biotic stress globally downregulates photosynthesis genes. Plant, Cell and Environment. 2010;33:1597-1613. DOI: 10.1111/j.1365-3040.2010.02167.x

[101] Puppe D, Sommer M. Experiments, Uptake Mechanisms, and Functioning of Silicon Foliar Fertilization—A Review Focusing on Maize, Rice, and Wheat. In Sparks DL Editor. Advances in Agronomy, v. 152; Academic Press, 2018. p. 1-49. https://doi.org/10.1016/bs.agron.2018.07.003.

Section 3

Abiotic Stress in Wheat

Chapter 7

Stresses in Plants: Biotic and Abiotic

Olayinka Bolaji Umar, Lawal Amudalat Ranti, Abdulbaki Shehu Abdulbaki, Abdulra'uf Lukman Bola, Abdulkareem Khadijat Abdulhamid, Murtadha Ramat Biola and Kayode Oluwagbenga Victor

Abstract

Plants are subjected to a variety of environmental stresses, which reduces and limits agricultural crop productivity. Environmental stresses that affect plants are of two types: biotic and abiotic stresses. Abiotic stress includes temperature, ultraviolet radiation, salinity, floods, drought, heavy metals, etc., which results in the loss of important crop plants globally, while biotic stress refers to damage caused by insects, herbivores, nematodes, fungi, bacteria, or weeds. Plants respond to all these environmental factors because the pants are fixed in a particular place. To cope with these stresses, a number of strategies have been developed by plants. They detect that the environmental stresses become activated and then generate the necessary cellular responses. Several investigations have been carried out to determine and understand plant assimilates partitioning and stress-tolerance plant genotype necessary for the understanding of the complexity of the response of a plant to biotic and abiotic stresses.

Keywords: biotic factors, environmental stresses, crop productivity, crop yield, tolerance mechanism

1. Introduction

Stress can be defined as any external and internal constraints that limit the photosynthetic rate and reduces the energy conversion ability of a plant to biomass [1]. Respond of a plant to stress is in different ways, some of which include variation in gene expression, cellular metabolism, growth rates, crop yields, and so on. Plant stress as a result of its response to varying environmental conditions. However, exposure to a particular stress by stress-tolerant plant species leads to the development of resistance with time to a particular stress [2]. The main types of stress that plants face are biotic and abiotic stresses. Abiotic stress is an environmental factor that is placed on plants, as a result of variation of physical or chemical stress [3], whereas biotic stress is a biological unit such as illnesses, insects, and other pests that are exposed to crop plants [4]. Some stresses cause injury in plants. These plants have a number of metabolic issues [5].

Plants can recover from injuries if the stress is light or only lasts a short time, as the effect is just transient; however, extreme stress results in death. However, many plants like xerophytic plants (Ephemerals) can escape the stress altogether. Biotic stress in plants is induced by living organisms, such as viruses, bacteria, fungus, nematodes, insects, arachnids, and weeds [2]. The agents that cause biotic stress deplete their hosts of nutrients, which can lead to plant mortality. Because of pre- and postharvest losses, biotic stress might become severe. Despite the absence of an adaptive immune system, plants have evolved sophisticated methods to deal with biotic stresses [6]. These stresses are controlled by the plant's genetic codes. Hence, there is a need to combat resistant varieties of crops so as to ensure food security and safety in subsequent growing seasons. Seed priming with growth and rooting hormones should also be considered.

2. Abiotic stresses and crop plants

Plants are subjected to a variety of abiotic stresses, all of which have an impact on crop yield around the world. The major biotic and abiotic stresses in plants are described in **Figure 1**. These include drought, salt, cold, heat, and toxins.

2.1 Drought

Water scarcity is a significant environmental limitation on plant productivity. Drought-induced crop output losses are likely to outnumber losses from all other sources because both the severity and duration of the stress are crucial [8].

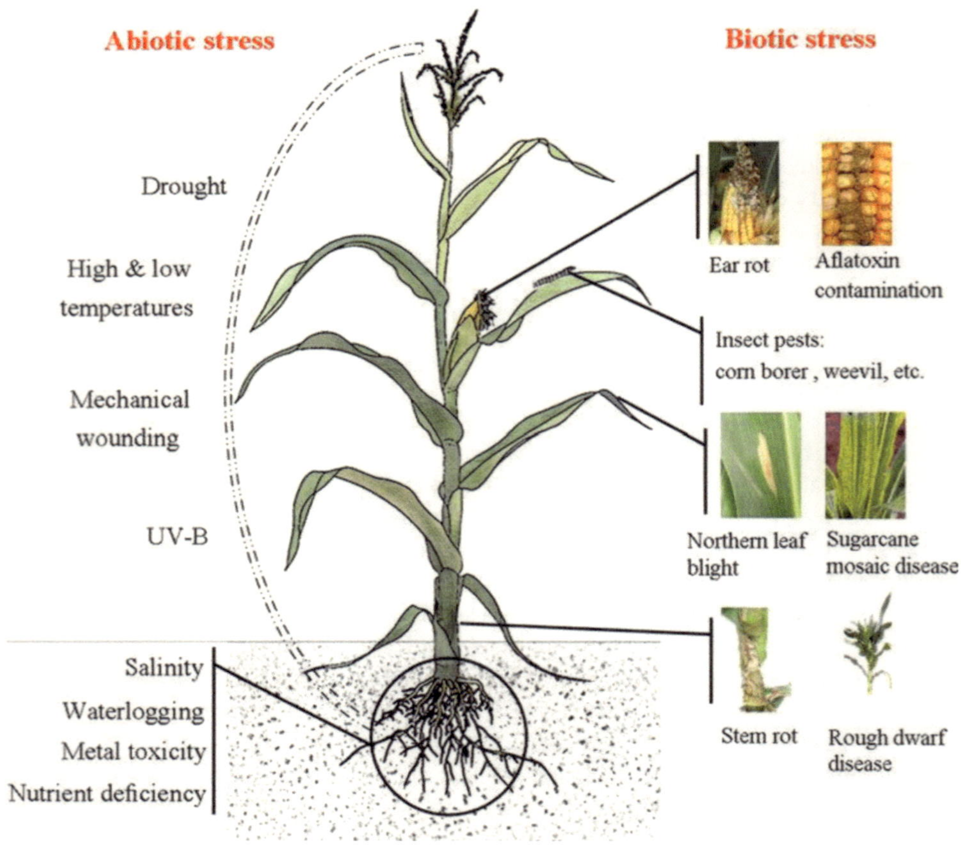

Figure 1.
An overview of major abiotic and biotic stresses [7].

The severity of the drought depends on the occurrence and distribution of rainfall, evaporative demands, and moisture storing capacity of soils, all of which are unpredictable [9]. Nowadays, climate has changed all around the globe by continuously increasing in temperatures and atmospheric CO_2 levels. The distribution of rainfall is unequal as a result of climate change, which functions as a major stress in the form of drought. Due to extreme drought conditions, the amount of soil water available to plants is steadily decreasing, causing plants to die prematurely. After drought is imposed on crop plants, growth will be arrested. Drought circumstances cause plants to lower their shoot growth, as well as their metabolic demands [7].

2.2 Salt

One of the most important limiting factors for crop growth and productivity is salt stress. Soil salinity is a global danger to world agriculture because it reduces crop yields and, as a result, crop productivity in salt-affected areas. Salinity is caused by the accumulation of salts in the soil or groundwater over a lengthy period of time as a result of natural processes or through human activities, for example, wethering of rocks or as a result of irrigation schemes using salt-rich irrigation water or having insufficient drainage [10]. There are several ways by which salt stress reduces the growth and yield of crops. Salt stress has two main effects on crop plants: osmotic stress and ion toxicity. These primary effects of salinity stress cause some secondary effects such as assimilate production, reduced cell expansion, and membrane function as well as decreased cytosolic metabolism [2].

2.3 Cold

Cold stress, as an abiotic stress, has been shown to be one of the most important abiotic stresses that reduce agricultural crop output by altering crop quality and post-harvest life. Many crop plant species have been found to be substantially hampered in their reproductive growth by chilling such as rice displaying sterility when exposed to chilling temperatures during anthesis [11]. Plants are sessile in nature; therefore, they have evolved unique ways to cope with temperature variations in their habitat [12]. In temperate conditions, plants are encountered by chilling and freezing conditions that are very harmful to plants as stress.

In order to adapt themselves, plants acquire chilling and freezing tolerance against such lethal cold stresses by a process called acclimation [13]. However, many important crops are still incompetent to the process of cold acclimation.

2.4 Heat

The temperature rises around the world have become a major problem, affecting not only plant development but also plant productivity, particularly in agricultural products. Heat stress has become the most important limiting factor to crop productivity and ultimately the food security [14]. When plants are subjected to heat stress, their seed germination rate, photosynthetic efficiency, and yield all suffer. Under heat stress, during the reproductive growth period, the function of a petal cell is lost, and the anther is dysplastic. For example, maize yields decrease sharply when the plants are exposed to temperatures greater than approximately 29–30°C [15].

2.5 Toxin

Toxic metals have been added to agriculture soils as a result of increased reliance on chemical fertilizers and sewage wastewater irrigation, as well as increasing industrialization, having detrimental consequences on the soil–plant environment system [16]. These metals bioaccumulate and slowly enter plants through air, water, and progression of the food chain over a certain period of time [17].

3. Crop plants and biotic stresses

Plants are subjected to a variety of biotic stress caused by various living organisms such as fungi, viruses, bacteria, nematodes, and insects [2]. These biotic stress agents induce a variety of diseases, infections, and damage to crop plants, lowering agricultural yields. However, different strategies for overcoming biotic stressors have been created through research methodologies. The biotic stresses in plants can be overcome by studying the genetic mechanism of the agents causing these stresses [18]. Genetically modified plants have proven to be a great effort against biotic stresses in plants by developing resistant varieties of crop plants.

Plant-parasitic nematodes feed on the contents of plant cells and can feed on all sections of the plant, but they predominantly cause soil-borne illnesses and affect the root system. They cause wilting and stunting, which are signs of nutritional inadequacy. Viruses cause not only local but also systemic damage to plants, causing stunting, chlorosis, and deformities in many areas of the plant, despite the fact that they rarely kill their hosts [19]. Plants are harmed when insects feed or lay eggs on them. Viruses can be transmitted to plants by piercing-sucking insects *via* their stylets. There are two types of fungus parasites: nectrotrophs, which use toxins to kill host cells, and biotrophs, which do not. They induce vascular wilts, leaf spots, and cankers, among other symptoms, and can infect different sections of the plant when combined with bacteria [20].

4. Plant defenses against abiotic stresses

Plants use five general botanical defenses against abiotic stresses. These include cuticle, unsaturated fatty acids, reactive species scavengers, molecular chaperones, and compatible solute, which are also an economic important trait [21].

4.1 Cuticle

This is the exterior translucent lipid structure in land plants, which seals the aerial surface of their organs. It is coated by cuticular waxes and is described as a hydrophobic layer. As the primary interface between plant and environment, the cuticle plays critical role in restricting liquid and gas fluxes, defending pathogen and insect attacks, and resisting various abiotic stresses. It is an elegant innovation of land plants to deploy an outermost shield derived from simple molecules, which is fundamental to their success in terrestrial colonization [22]. Wax accumulation in the cuticle is closely associated with multiple stress tolerance [23].

4.2 Unsaturated fatty acids

Unsaturated fatty acids containing 16 or 18 carbon atoms are the key ingredients of the membrane and the prime stocks for the cuticle. The unsaturated nature of fatty acids is a major determinant of membrane fluidity [21]. Dysfunction of biomembrane due to protein deactivation and ion leakage are caused by cold-driven rigidification and heat-driven fluidization, which makes membrane fluidity susceptible to various abiotic stresses, especially at high temperatures [24]. An increase in the level of normal alkanes with a decrease in the level of primary alcohols can lead to cold susceptibility, which can cause growth retardation, while an increase in the levels of both n-alkanes and primary alcohols resulted in better viability, where drought and freezing will have no effect on plant growth [25]. When polyunsaturated fatty acids are liberated by lipase form glycerolipids, they serve as raw materials for the synthesis of oxylipins, a bioactive molecule that is involved in the diverse physiological processes of stress resistance [26].

4.3 Reactive species scavengers

The reactive species scavengers include reactive carbonyl species (ROS) and reactive oxygen species (RCS). The ROS and RCS are interwoven, due to the fact that RCS can arise from ROS-induced lipid peroxidation, while ROS can be raised by RCS activities the other way round. Abiotic stresses can trigger a burst in both ROS and RCS thereby turning the two scavengers into a general defenses. Plants utilize both enzymatic and non-enzymatic means to developed sophisticated ROS scavenging system [21]. The application of excessive nitrogen fertilization in crop cultivation depresses the ROS scavenging system causing the increase in stress susceptibility [27].

4.4 Molecular chaperones

Molecular chaperones are induced to facilitate protein folding, assembly, transport, and degradation. Heat shock protein (HSP), which are good examples of molecular chaperon, is employed by all living organisms to counteract all detrimental conditions that can induce protein damage, wherein they function to prevent aggregation of denatured proteins, assist in their refolding, or present them to lysosomes or proteasomes for proteolysis, thereby restoring cellular homeostasis [28].

4.5 Compatible solutes

They are electrical neutral small organic compounds with high solubility and low toxicity. The molecules include sugar, amino acids, and their derivatives [21]. In an abiotic stress, these metabolites may accrue to act as osmoprotectants against dehydration, scavengers of RS, and/or stabilizers of proteins and membranes [29].

5. Conclusion

Plants are sessile organisms that are susceptible to environmental damages. In a broad sense, both biotic (viruses, bacteria, insects) and abiotic (heat, drought, salt, etc.) are adversaries facing world food production. Plants affected by these biotic and abiotic stress factors surfers physiological and metabolism changes. Hormonal and genetic defense mechanisms of the plant are also affected. Here, there is a need for phytologist and plant Breeders to develop tolerant varieties so as to combat these stresses to ensure good security. Plants will continue to be subjected to biotic and abiotic stresses until responsive mechanisms are created, and this will pose a significant threat to global agriculture. In plant cells, glycolysis operates as the principal source of this cytotoxin, due to the non-enzymatic dephosphorylation of two intermediates, glyceraldehyde 3-phosphate and dihydroxyacetone phosphate. Once over accumulated, methylglyoxal can also damage various biomolecules, especially with its aldehyde group.

Author details

Olayinka Bolaji Umar[1*], Lawal Amudalat Ranti[2], Abdulbaki Shehu Abdulbaki[3,4], Abdulra'uf Lukman Bola[5], Abdulkareem Khadijat Abdulhamid[1], Murtadha Ramat Biola[1] and Kayode Oluwagbenga Victor[1]

1 Faculty of Life Sciences, Department of Plant Biology, University of Ilorin, Ilorin, Nigeria

2 Department of Biology, School of Sciences, Kwara State College of Education, Ilorin, Nigeria

3 Faculty of Life Sciences, Department of Biological Sciences, Federal University Dutsinma, Katsina State, Nigeria

4 Faculty of Sciences, Department of Biology, King Abdulaziz University Jeddah, Saudi Arabia

5 Faculty of Pure and Applied Sciences, Department of Chemistry, Kwara State University, Malete, Nigeria

*Address all correspondence to: olayinka.bu@unilorin.edu.ng

References

[1] Atafar Z, Mesdaghinia A, Nouri J, Homaee M, Yunesian M, Ahmadi M, et al. Effect of fertilizer application on soil heavy metal concentration. Environmental Monitoring and Assessment. 2009;**160**:83-89

[2] Fahad S, Bajwa AA, Nazir U, Anjum SA, Farooq A, Zohaib A, et al. Crop production under drought and heat stress: plant responses and management options. Frontiers in Plant Science. 2017;**29**:01147

[3] Fich EA, Segerson NA, Rose JK. The plant polyester cutin: Biosynthesis, structure, and biological roles. Annual Review of Plant Biology. 2016;**67**: 207-233

[4] Gimenez E, Salinas M, Manzano-Agugliaro F. Worldwide research on plant defense against biotic stresses as improvement for sustainable agriculture. Sustainability. 2018;**10**:391

[5] Godoy F, Olivis Hernadez K, Stange C, Handford M. Abiotic stree in crop science: Improving tolerance by applying plant metabolites. Plants. 2021;**10**:186

[6] Gong F, Yang L, Tai F, Hu X, Wang W. "Omics" of maize stress response for sustainable food production: Opportunities and challenges. OMICS: A Journal of Integrative Biology 2014;**18**(12):714-732

[7] Grime JP. Evidence for the existence of three primary strategies in plants and its relevance to ecological and evolutionary theory. The American Naturalist. 1977;**111**:1169-1194

[8] Gull A, Lone AA, Wani NUI. Biotic and abiotic stresses in plants. In: de Oliveira AB, editor. Abiotic and Biotic Stress in Plants. IntechOpen; 2019

[9] Hazel JR. Thermal adaptation in biological membranes: Is homeoviscous adaptation the explanation? Annual Review of Physiology. 1995;**57**:19-42

[10] He M, He CQ, Ding NZ. Abiotic stresses: General defenses of land plants and chances for engineering multistress tolerance. Frontiers in Plant Science. 2018;**19**:1171

[11] Jiang W, Lee J, Chu SH, Ham TH, Woo MO, Cho YI, Koh HJ. Genotype × environment interactions for chilling tolerance of rice recombinant inbred lines under different low temperature environments. Field Crops Research 2010;**117**(2-3):226-236

[12] Jones RAC. Global plant virus disease pandemics and epidemics. Plants. 2021;**10**:233

[13] Kong LA, Xie Y, Hu L, Si JS, Wang ZS. Excessive nitrogen application dampens antioxidant capacity and grain filling in wheat as revealed by metabolic and physiological analyses. Scientific Reports. 2017;**7**:43363

[14] Liliane TN, Charles MS. Factors affecting yield of crops. In: Amanullah, editor. Agronomy – Climate Change & Food Security. IntechOpen; 2020. DOI: 10.5772/intechopen.90672 Available from: https://www.intechopen.com/chapters/70658

[15] Liu X, Zhou Y, Xiao J, Bao F. Effects of chilling on the structure, function and development of chloroplasts. Frontiers in Plant Science. 2018;**22**:1715

[16] Nievola CC, Carvalho CP, Carvalho V, Rodrigues E. Rapid response of plants to temperature changes. Temperature. 2017;**4**(4): 371-405

[17] Rauwane M, Ntushelo K. Understanding biotic stress and hormone signalling in cassava (*Manhot esculenta*): Potential for using

hyphenated analytical techniques. Applied Sciences. 2020;**20**:8152. DOI: 10.3390/app10228152

[18] Rollins JA, Habte E, Templer SE, Colby T, Schmidt J, von Korff M. Leaf proteome alterations in the context of physiological and morphological responses to drought and heat stress in barley (*Hordeum vulgare* L.). Journal of Experimental Botany. 2013;**64**(11): 3201-3212

[19] Savchenko TV, Zastrijnaja OM, Klimov VV. Oxylipins and plant abiotic stress resistance. Biochemistry. 2014;**79**:362-375

[20] Schlenker W, Roberts MJ. Nonlinear temperature effects indicate severe damages to U.S. crop yields under climate change. Proceedings of the National Academy of Sciences of the United States of America. 2009;**106**: 15594-15598

[21] Slama I, Abdelly C, Bouchereau A, Flowers T, Savoure A. Diversity, distribution and roles of osmoprotective compounds accumulated in halophytes under abiotic stress. Annals of Botany. 2015;**115**:433-447

[22] Srivastava V, Sarkar A, Singh S, Singh P, de Araujo ASF, Singh RP. Agroecological responses of heavy metal pollution with special emphasis on soil health and plant performance. Frontiers in Environmental and Environmental Toxicology. 2017;**3389**:00064

[23] Visser EJW, Voesenek LACJ, Vartapetian BB, Jackson MB. Flooding and plant growth. Annals of Botany. 2003;**91**:107-109

[24] Wang W, Vinocur B, Shoseyov O, Altman A. Role of plant heat shock proteins and molecular chaperones in the abiotic stress response. Trends in Plant Science. 2004;**9**:244-252

[25] Wery J, Silim SN, Knights EJ, Malhotra RS, Cousin R. Screening techniques and sources and tolerance to extremes of moisture and air temperature in cool season food legumes. Euphytica. 1994;**73**:73-83

[26] Xue D, Zhang X, Lu X, Chen G, Chen ZH. Molecular and evolutionary mechanisms of cuticular wax for plant drought tolerance. Frontiers in Plant Science. 2017;**8**:621

[27] Yadeta KA, Thomma BPHJ. The xylem as battleground for plants hosts and vascular wilt pathogens. Frontiers in Plant Science. 2013;**23**:97

[28] Zaid H, Tamrakar AK, Razzaque MS, Efferth T. Diabetes and metabolism disorders medicinal plants: A glance at the past and a look to the future 2018. Evidence-based Complementary and Alternative Medicine. 2018:5843298

[29] Zhang JY, Broeckling CD, Sumner LW, Wang ZY. Heterologous expression of two *Medicago truncatula* putative ERF transcription factor genes, WXP1 and WXP2, in *Arabidopsis* led to increased leaf wax accumulation and improved drought tolerance, but differential response in freezing tolerance. Plant Molecular Biology. 2007;**64**:265-227

Chapter 8

Drought Affected Wheat Production in Bangladesh and Breeding Strategies for Drought Tolerance

Afsana Hannan, Md. Najmol Hoque, Lutful Hassan and Arif Hasan Khan Robin

Abstract

Wheat is one of the major cereal crops in Bangladesh. Over the last two decades, wheat consumption has passionately amplified in Bangladesh but its production has declined due to various stress environments. Recurrent drought event due to climate change that threatens the country's food safety has become a serious concern. To safeguard the food security, adopting suitable breeding strategies can add momentum. Developing drought tolerant wheat varieties are the definitive means of protecting the crop against hostile effects of drought. Plant breeders are exploring various breeding strategies to breed for the varieties that can cope with water deficient conditions well. Besides, breeders are consistently looking for new prospects and strategies that can boost genetic gain in yield. To endorse drought tolerance in wheat, understanding the physiological and genetic adaptation mechanisms of wheat cultivars during drought stress would provide the estimated benchmarks to adjust for suitable breeding programs. The efforts of developing drought tolerant wheat genotypes could be supported by different breeding strategies including *in vitro* haploid and double haploid protocols, polyploidization, development of various types of hybrids and induced mutants by utilizing both classical and molecular breeding techniques. The proposed book chapter shall discuss the pattern of drought-stress in the wheat growing regions, effects of drought stress on wheat production and suitable breeding strategies for developing drought tolerant genotypes in Bangladesh.

Keywords: wheat breeding, drought stress, tolerance mechanisms, breeding strategies

1. Introduction

Bangladesh is a small country geographically situated in between Himalaya and Bay of Bengal. It is among the most vulnerable countries in world to future climate change due to the flat deltaic topography, very low elevation (below 10 meters above sea level) and high population density [1–3]. Eating a lot of rice is the primary food habit of Bangladeshi people. Next to rice, wheat is the second most important cereal crops in Bangladesh for attaining food and nutritional security [4]. Although being one of the major rice producers and consumers in world [5], consumption and import of wheat in Bangladesh are growing significantly over the years [5, 6–9]. The speedy economic growth, swift urbanization, and the associated alterations in lifestyle are accountable for the increased consumption of wheat which is not going to change [8]. Instead the demand of wheat will be enhanced in near future [4]. Despite increasing yield, gradual decrease of wheat growing area make the domestic wheat production curve more or less static [10]. At present, the domestic production of the country can only encounter around 20% of total wheat consumption [11, 12] and import is the only way for meeting her demand–supply gap [6]. Several periodic natural calamities such as salinity, drought, high temperature stress, flash floods and cyclones have been accelerated due to climate change in recent years [4, 13]. Among the abiotic stresses, drought is the most prominent and prevalent limiting factors of wheat production [14–16]. Rising temperature and changing in precipitation pattern lead to increasing incidence and intensity of drought events in country like Bangladesh [17–21]. Drought employs expressively adverse effects on production of winter crop wheat in northern and central part of Bangladesh [22, 23]. Around 3.5 million ha land are vulnerable to crop production due to drought and wheat is one of the major cereal crops under the radar of this threat [24]. Considering these facts, drought should be highly preferred in future wheat improvement programs. For attaining self-sufficiency in wheat production, wheat breeders of Bangladesh have no alternatives but to develop well adapted drought tolerant varieties [22]. In spite of the polygenic nature, there are ample opportunities to increase drought tolerance of wheat through making some alterations in genetic and molecular levels. Therefore recent wheat breeding programs for drought stress should focus on utilization of both conventional as well as advanced molecular techniques.

2. Pattern and distribution of drought stress in Bangladesh

In Bangladesh, drought is defined as the period when soil moisture content is less than the required amount for satisfactory growth of a crop during a normal crop growing season [25]. According to assessment of Intergovernmental Panel on Climate Change (IPCC), by the year of 2050 about 8 million people of Bangladesh will be affected by droughts [26]. Due to tropical humid type climate, Bangladesh faces widely varying seasonal rainfall pattern, moderately warm temperatures and high humidity [27]. Irregular and varying rainfall pattern due to climate change and lack of surface water is the main reasons of recurrent devastating drought events in many areas of Bangladesh [28, 29]. Among the meteorological droughts, seasonal drought due to asymmetrical distribution of standard rainy and dry season and contingent drought due to irregular rainfall are more predominant in Bangladesh [25]. Due to high variability in pattern and distribution of rainfall, the north-western part of Bangladesh become more susceptible to droughts [30, 31]. In addition, groundwater resources are continuously abused by the farming communities causing scarcity in surface water [32, 33]. Over the last 2–3 decades, the northwestern part of Bangladesh (Barind tract) has been more exposed to recurrent drought events than the other parts [34]. Majority of the parts of greater Dinajpur, Rangpur, Pabna, Rajshahi, Bogura, Naogaon and Joypurhat districts are included in Barind Tract shown in **Figure 1** characterized by relatively less rainfall (average annual rainfall 1329 mm), shortage of surface water and high temperatures [25, 35]. One of the most vulnerable districts to droughts in Bangladesh is Rangpur [36].

DOI: http://dx.doi.org/10.5772/intechopen.95283

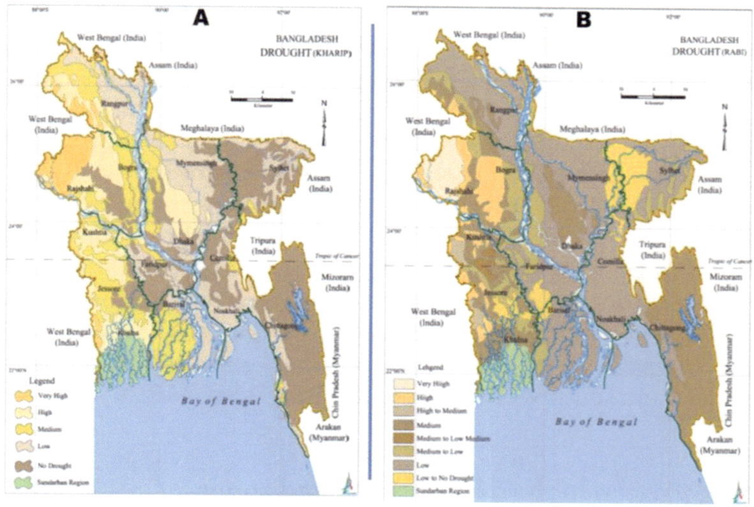

Figure 1.
Map of Bangladesh showing drought prone areas A. in kharif season B. in Rabi season [25].

Because of the extreme climate fluctuations mainly in the patterns of rainfall, Bangladesh is predicted to face increased rainfall upto 5–6% by 2030 resulting prolong flood during monsoon season and severe drought outside the monsoon season [13, 34]. Inadequate pre-monsoon shower, a delay in inception of rainy season or a quick advent of the monsoon season may accelerate the drought risk in Bangladesh [37]. Bangladesh experienced 20 different drought events over the last 50 years and among them the droughts of 1973–1974, 1975, 1978–1979, 1981, 1982, 1989, 1994–1995, 2000 and 2006 are most hazardous [34, 38]. Effects of some major historical drought events of Bangladesh are presented in **Table 1**.

In Bangladesh, the spatial pattern of pre-monsoon droughts are more recurrent in northwestern part [39]. An analysis on monthly pattern drought from 1971 to 2010 has suggested that Dinajpur, Kushtia, Rajshahi, Rangpur and Bogura are the highest drought-prone parts of the country [40]. Further drought trends investigation has revealed the declining trends in rainfall and increase in dryness at Ishurdi, Bogura, Sayedpur and Rangpur [41]. Investigation on spatiotemporal drought

Happening year	Drought impacts
1973–1974	One of the most severe drought events in the century that caused famine in 1974 in northern part of Bangladesh
1975	Affected 47% of area and half of the total population of Bangladesh
1978–1979	Affected about 42% of the cultivated land and 44% of the total population. Caused severe damage to crop production especially rice (reduced about 2 million tons production)
1981	Adversely affected crop production
1982	Caused severe reduction in rice production (reduced about 53,000 tons production)
1989	Dried up most of the rivers in north-western regions of Bangladesh with dust storms in Nawabganj, Naogaon, Nilpahamari and Thakurgaon districts
1994–1995 and 1995–1996	Caused immense crop damage, especially to the main crops of northwest Bangladesh like rice, jute and bamboo clumps. The most persistent droughts in recent times

Table 1.
Major historical droughts and its impact in Bangladesh [28].

patterns on a regional scale has exposed that higher intensities and frequencies of drought events in the northwestern part make the area more vulnerable to both drought severity and extremity [42]. Recent assessment of droughts from 1960 to 2011 in context of changing climate using drought hazard index (DHI) and drought index (DI) has disclosed that the northern part of Bangladesh are more drought-prone and there is a probability for the area of experiencing more extreme drought events in near future [43]. The studies on changing pattern of meteorological droughts indicates the rising trend of more extreme droughts in cropping season and also reveals the possibility of changing the drought occurrence pattern in both areas where it historically affected most (northwestern part) or the areas with fewer droughts (other parts) [44, 45]. Huge uncertainties are noticed in the possible future changes in droughts and also that would expand from north-western to central, western and south-western regions in Bangladesh [46, 47].

3. Cropping pattern in the drought-affected zones

Cropping pattern of an area is normally determined by its climatic parameters related to a particular time of a year. Bangladesh is situated in subtropical region giving it a suitable temperature range which makes it favorable for year round crop cultivation. However, Bangladesh has a complex and intensively diverse cropping pattern and that pattern is evolving and changing at a continuous basis [48]. Depending on cultural method, the whole crop-growing period of Bangladesh is distributed into two major seasons i.e. Kharif season and Rabi season. Beside these two, there is a transitional season named pre-kharif (shown in **Table 2**) [49]. Kharif crops like rice, jute, maize, millets etc. are grown in Kharif season and Rabi crops like wheat, mustard, chickpea, lentil etc. are grown during Rabi season [25].

In Bangladesh, all the cropping season are more or less affected by drought. But pre-monsoon and post-monsoon period are mostly prone to drought events [25]. Kharif drought negatively affects the critical reproductive stage of transplanted Aman rice where all of the Rabi crops are affected by pre-kharif/rabi droughts [4]. Assessment of drought in northern area of Bangladesh for the period between 1971 and 2008 reveals that most extreme drought conditions have been experienced in Rabi season including pre-monsoon [24]. Increasing trend in precipitation change in Bangladesh causes more rainfall in monsoon and less rainfall in winter resulting in droughts in winter season. Thus yield of various crops like HYV boro rice, aus rice, wheat, sugarcane, pulses and potatoes growing in Rabi and pre-kharif season are badly affected by droughts [35, 50]. In recent decades, the drought condition in northwestern Bangladesh severely affected the production of rice and all Rabi crops

Cropping season	Occurring month	Characteristics
Kharif or Monsoon (also known as kharif-2 or aman)	June/July to September/ October	• High rainfall, temperature and humidity • Enough moister in soil • Rain-fed crops are grown
Rabi or Winter	October/November to February/March	• Little or no rainfall during the season • Crops grown under irrigation
Pre-kharif or pre-monsoon or spring (also known as kharif-1)	March/April to May/June	• Unreliable rainfall • Intermittent moisture supply to crops

Table 2.
Major cropping seasons in Bangladesh.

(wheat, tobacco, sugarcane etc.) [25]. Rice-rice, rice-wheat and rice-maize are the dominating cropping patterns in Bangladesh in the drought regions [51, 52]. In late October to early November, certain areas of lands in Bangladesh become empty because of using short duration rice varieties which is appropriate for wheat cultivation [4]. For decades, wheat is grown in wheat-fallow-T. aman rice cropping pattern in north-western part of Bangladesh with some exceptions like wheat-jute-T. aman rice cropping pattern [53].

4. Adverse effects of drought on wheat production

Drought is one of the most limiting stress factors for crop growth and development, dry matter production and potential yield [15, 54]. The major processes required for plant growth and development are hampered by the drought condition. Water deficit conditions lessen the rate of photosynthesis by inhibiting chlorophyll synthesis, impede cellular elongation and metabolism, decline the CO_2 assimilation rates due to reduction in stomatal conductance and gaseous exchange, reduce dry matter biomass production and alter root morphology [54, 55]. As a result leaf size, stem elongation, root production and finally the rate of growth and yield are affected by drought [54].

Drought is not a static stress, it can occur at any crop growing period, its severity and frequency can vary and also it can recurrently happens in combination with other abiotic stresses, such as salinity and heat [56]. Drought stress can fluctuate diurnally (high during peak photosynthetic period and low overnight) and different organs of plants respond differently to drought stress [57]. Yield contributing traits vary according to growth stage of plant, so the level of seriousness of drought stress eventually relies on the particular growth stages that are impacted by drought. The nature of plants' response also differ depending on whether the plant is experiencing stress for the first time or after several exposures and whether they are recovering from stress after a rainfall or irrigation event [58].

Water is needed for the entire growth period of wheat but some specific stages are more sensitive to water limitations. Various morphological, physiological and biochemical alterations are occurred in plants body under drought environment (see **Table 3**). In case of wheat, the extent of drought stress may vary according to different growth stages. Specific critical growth stages of wheat plants such as germination and seedling stages [60]; tillering and stem elongation stages [61, 62]; heading, anthesis and grain filling stages [16, 60] may be more vulnerable

Drought stress in wheat	Morphological alterations	Limited plant size, ceased plant height, reduced leaf extension, lessened leaf size and number of leaves, decreased leaf area, reduced leaf longevity, prompt maturity, augmented root-to-shoot ratio, condensed total shoot length, lowered yield
	Physiological alterations	Stomata closure, reduction in photosynthesis, swift in oxidative stress, alterations in cell wall integrity, decrease in leaf water potential, lessen growth rates, reduced transpiration rates and relative water content, developed water use efficiency
	Biochemical alterations	Reduction in rubisco efficiency, decrease in photochemical efficiency, production of reactive oxygen species (ROS), increase in oxidation damage, hampered antioxidant defense system, reduced chlorophyll content

Table 3.
Effect of water-deficit stress on morphological, physiological and biochemical traits of wheat [59].

to drought stress. Long term droughts (starting from stem elongation through to maturity) cause more drastic yield reduction compared to those initiating at later stages through to maturity [63]. Although the influence of drought stress on heading and grain filling stages are more severe in terms of yield, drought can also negatively affect the multiple growth stages of wheat comprising germination, tillering, booting, heading, anthesis, and maturity [64].

5. Crop traits and mechanisms adaptive to drought stress

In drought condition, sometimes very swift growth responses are generated even due to a little water pulse that can vigorously activate plant growth and safeguard survival [65]. Constantly fluctuating nature of drought events makes it indispensible to understand the plants' aptitude to adapt and recover from the stress [66]. To overcome the harmful effects of drought stress, naturally plants are well furnished with various adaptive mechanisms. These adaptive mechanisms support plants for an optimal maintenance of growth for metabolic regulation and survival [67]. The more the extent of these mechanisms, the more will be the plants' capability to overcome stress condition. But the adaptive mechanism is not as simple as it sounds; it comprises diverse morphological, physiological and anatomical modifications in plant under stress condition. Morphological and metabolic adaptation processes of plants vary according to cultivars in response to water deficit condition. As, plants' may have unique adaptation capabilities irrespective to cultivars [68]. Different physiological processes in plant such as photosynthesis, heat dissipation and chlorophyll fluorescence are occurred in rivalry with each other in response to drought events i.e. any upsurge in the efficacy of one will bring diminution to others [69]. The normal fluctuation values of these physiological processes can denote plant fitness with the magnitude of environmental stress [55].

Drought adaptation is complicated that experiences diverse anatomical and morpho-physiological and biochemical amendments in plants such as alterations in leaf traits or canopy cover, leaf water relations with modification of growth rates, reduction of stomatal opening and associated components [70, 71]. The plants' response to water deficit condition has been extensively studied to recognize tolerance mechanisms [72]. So, detailed knowledge about underlying behavior of plants under drought stress is required to develop drought tolerant plants. Although being complex, mainly three kinds of drought-resistance mechanisms are exhibited by plants to evade the resulting devastating effects of droughts: (i) drought escape (ii) drought avoidance and (iii) drought tolerance [73]. Drought escape happens when plants grow quickly and reproduce before severe drought conditions. In this mechanism, plants evades drought season by modifying flowering time thus they try to complete their life cycle before drought condition. In drought avoidance mechanism, plants avoid water-deficit situation by enhancing their water-use efficiency (WUE) through closure of stomata, reduction of transpiration, limitation of vegetative growth, or by increment in root growth. In case of drought tolerance, drought stress is fought by plants at cellular level through osmotic adjustment by developing antioxidants and production of molecules that stabilize proteins [73].

Wheat plants exhibit a tight network of morpho-physiological and photoprotective mechanisms to alleviate the drought stress [66]. To escape reproductive failure from severe drought stress, plants displayed phenological alterations of earlier anthesis and maturity [66]. Previous literature revealed constitutive traits that confer dehydration avoidance mechanisms in plants include leaf waxy layer, leaf rolling and osmotic adjustment [74], high root length density [75] and high fine roots with small diameters [76], a deep root system and the number of seminal roots [77–79],

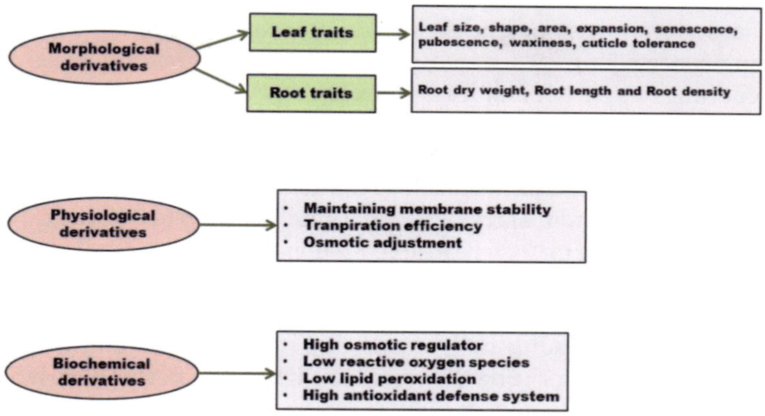

Figure 2.
Morpho-physiological and biochemical derivatives of drought tolerance in wheat.

high total root length and total root surface area [80, 81], root-to-shoot dry matter ratio [82] and root partitioning of assimilates to shallow or depth roots in response to drought [83]. There are range of morphological, physiological and biochemical derivatives of drought tolerance in wheat [59] (shown in **Figure 2**). Water use, water use efficiency, biomass yield and flag leaf relative water contents are the important drought tolerance traits in wheat [84, 85]. Selection of wheat plant with high transpiration efficiency, high percentage of relative water contents and cell membrane thermo-stability and greater osmotic adjustment capacity leads to produce drought tolerant plants [59, 86]. When drought stress is imposed on seedling stage of wheat, cell membrane thermo-stability, fresh and dry weight of seedlings are considered major traits to govern drought responses under stress conditions [87]. Therefore, greater morphological adaptation with limited down-regulated physiological activities followed by high recovery in wheat cultivars designate its capability to effectively endure drought events [66].

6. Prospects of breeding for drought tolerance in wheat

Plant breeders around the world have to deal with great challenges to work with drought stress. Polygenic nature of drought makes the breeding efforts more complicated than other abiotic stresses [88]. Global climate change will result in frequent drought events as per predicted in country like Bangladesh [89]. So, for improving wheat production in Bangladesh, research priority should be focused on breeding new high yielding drought tolerant wheat varieties. Majority of the studies under drought stress focus on the response of natural drought in field conditions where drought events are ambiguous and irregular using conventional techniques. Generally the conventional breeding techniques such as introduction, selection, hybridization and mutation are being used by the breeders of Bangladesh. Whereas throughout the globe, wheat breeders are now using different novel breeding methods including *in situ* and *in vitro* techniques. Under drought stress, several morpho-physiological and biochemical mechanisms are activated in plant body to withstand the stress. But poor conceptual knowledge about the developmental and physiological basis of yield related traits under water-deficit environments make the drought stress more complex [90]. Therefore, better understanding about the detailed physiological and genetic adaptive strategies of wheat cultivars during water-deficit stress would offer the appraised benchmarks of breeding methods for pursuing drought tolerance in wheat [59]. Hence, selection procedures based on physiological traits have potentiality to improve the final productivity of wheat under drought stress [66].

In recent times, as part of empirical breeding based programs, breeders have been embracing replicated, multi-locational and multi-year variety testing for finding out the best adaptive varieties to stress environments. Expanding grain yield under drought stress can be performed to a limited extent through selection process [91, 92]. For being recurrent and season indefinite stress event, trait evaluation under drought condition may cause losing of potential genetic resources which perform better in normal wheat-growing environments [89]. This may ultimately hamper the variety development process. Therefore, evaluation including diverse testing environments including both normal and stressed conditions will be more suitable and competent for the development of high yielding, stable varieties amended to water-deficit conditions [89].

7. Breeding strategies for drought tolerance in wheat

It is very challenging for the plant breeders of Bangladesh to develop drought-tolerant wheat varieties [22]. For ensuring future food security of Bangladesh, the scientists of Wheat Research Center (WRC) of Bangladesh Agricultural Research Institution (BARI) are trying hard to develop wheat varieties that can be suited well in abiotic stress environments [4]. But alongside using a range of conventional breeding strategies for developing stress tolerant variety, breeders always search to produce new genetic variant to increase of genetic gain through advanced molecular approaches.

For maintaining the consistency of wheat production in Bangladesh adaptive to future climate change, the wheat varieties of next generation should possess high yield potentially even under stressed conditions. Yield potentiality can be enhanced through strategic crosses depending upon pyramiding yield potential traits and related physiological traits to stress tolerance in well adapted genotypes [4]. Breeding for drought tolerance in wheat initially requires satisfactory amount of variability among the source populations. Conventional hybridization is the most widely used breeding procedure in wheat, where genetic variability is created through combination and recombination of desirable genes in the background of diverse adapted genotypes followed by a selection of desirable plants in subsequent generations to develop improved varieties for the target environment [4]. Generally grain yield is the primary basis for selection for drought tolerance but indirect selection based on related yield-contributing and physiological traits can be more effective for developing drought tolerant varieties [89, 93–95]. In this connection, several wheat lines collected from various national and international sources especially CIMMYT (International Maize and Wheat Improvement Center) are evaluated for their performance in diverse growing environments of Bangladesh [4]. Screening of drought tolerant wheat genotypes has been commenced at Barind area of Rajshahi region of Bangladesh where incorporation of related traits to drought tolerance into adapted varieties is also undergoing [4]. Although being the main breeding procedures with some advantages, conventional techniques are slow, labour-intensive and economically unfeasible [96].

In contrast to time-consuming conventional breeding methods for accomplishing homozygous lines to develop wheat varieties, double haploid breeding instantly enables development of homozygous lines from a crop plant. Hence, double haploid breeding can be also an effective method in wheat breeding since selection efficiency relies on uniform homozygous line production. But, unwanted genetic modifications due to gametoclonal variation negatively affect the selection of population [97–99]. Interspecific crosses can also produce double haploids of wheat. Recently, WRC of BARI (now, Bangladesh Wheat and Maize Research Institute) has embraced the double haploid breeding technique through cross-pollinating wheat and maize [4]. For speeding up the variety release process, scientists are being trained for efficient targeted crosses to produce double haploid plants [4]. Mutation breeding offers another way to produce drought tolerant wheat varieties in Bangladesh. Induced mutations by gamma-ray is very efficient in augmenting genetic variability which provide a great opportunity for the wheat breeders to select for drought tolerance in M_2 (mutant generation 2) and next mutated generations [100–102]. Recently, in bread wheat, drought tolerant mutants are formed using gamma rays that lead to the release of 26 varieties worldwide [103]. Incorporating with several improved traits, these varieties can survive the stress environments. Thus, high potentiality of developed wheat mutants for direct release and inclusion in hybridization breeding programs is the major benefit of mutation breeding [104].

Molecular mechanism of drought tolerance is very complicated to understand. Numerous drought-responsive genes are involved in making plant drought tolerant, furthermore expressions of these genes also differ with various plant growth stages [74, 105]. Various genes and their related enzymes and proteins including late embryogenesis abundant (lea), responsive to abscisic acid (Rab), rubisco, helicase, proline, dehydrins, vacuolar acid invertase, glutathione-S-transferase (GST) and carbohydrates provide the molecular basis for drought tolerance in wheat [59]. It points towards challenges and uncertainties remain in breeding for drought tolerance. Hence, inclusion of innovative molecular and biotechnological methods like molecular marker methods, quantitative trait loci (QTL) mapping strategies, expression patterns of genes and genetic engineering should be practiced for the development of drought tolerant wheat genotypes. Currently, molecular markers are extensively used for detecting the location of drought-induced genes. Genome mapping and tagging of various traits aided by molecular markers are utilized in Marker-assisted breeding in wheat for developing drought tolerance [106]. Marker techniques allow indirect selection independent of crop developmental stage specially when dealing with polygenic trait like drought tolerance. In the previous few decades, molecular markers like isozymes, SDS-protein and sequence based DNA markers are exploited in wheat breeding for assessing gene diversities, precise mapping of their respective QTLs on chromosomes and finally for selecting quantitative traits like drought tolerance [107–111]. Even though large genome size of wheat, polygenic nature of the trait, instability of some QTL ultimately make the mapping process very challenging to execute for drought tolerance [106, 112, 113].

Now-a-days, modern biotechnological approaches have been involved in developing transgenic plants that can withstand the severity caused by drought. Since, these biotechnological strategies enable more understanding about the drought responses of crops at the entire plant and molecular levels [114]. It is evident from previous study that in field conditions, genetically modified wheat exhibits high tolerance to drought [115]. Plant tissue culture, hydroponic culture, *in situ* techniques and *in vitro* techniques such as somaclonal variants selection, protoplast culture should be employed for breeding under drought stress [116]. Further novel technologies like genome editing [117], high throughput phenotyping (HTP) and next generation sequencing (NGS) may be employed to explore innovative possibilities for improving drought tolerance in wheat plants [89, 118–120].

8. Conclusions

As it is an urgent call for upgrading wheat production under increasing potentiality of drought events, wheat breeders of Bangladesh need to emphasize on integrating more breeding techniques to make drought tolerant varieties. Majority of the breeding approaches here are concentrating on conventional techniques. So, it is high time to combine the conventional breeding methods with the modern techniques to develop wheat genotypes for the next generation. New advanced screening, hybridization and selection techniques shall need to be incorporated with conventional techniques. To maximize the breeding efficiency for drought tolerance in wheat, advanced precision phenotyping accompanied by genetic and molecular approaches should be integrated in breeding programs.

Conflict of interest

"The authors declare no conflict of interest."

Funding

This research was supported by the University Grants Commission of Bangladesh (Grant No. 2019/829/UGC).

Author details

Afsana Hannan[1], Md. Najmol Hoque[2], Lutful Hassan[1] and Arif Hasan Khan Robin[1*]

1 Department of Genetics and Plant Breeding, Bangladesh Agricultural University, Mymensingh, Bangladesh

2 Department of Biochemistry and Molecular Biology, Khulna Agricultural University, Khulna, Bangladesh

*Address all correspondence to: gpb21bau@bau.edu.bd

References

[1] Ahmed AU. Bangladesh climate change impacts and vulnerability. A synthesis; climate change cell; department of environment, comprehensive disaster management programme: Dhaka, Bangladesh, 2006.

[2] Shahid S, Behrawan H. Drought risk assessment in the western part of Bangladesh. Natural Hazards. 2008; 46:391-413. DOI: 10.1007/s11069-007-9191-5

[3] World Bank Bangladesh – country assessment strategy FY 2011-2014. Bangladesh Country Management Unit, South Asia Region, The World Bank Office, Dhaka, 2009.

[4] Barma NCD, Hossain A, Hakim MA, Mottaleb KA, Alam MA, Reza MM, Rohman MM. Progress and challenges of wheat production in the era of climate change: a Bangladesh perspective. In Wheat Production in Changing Environments. Springer, Singapore. 2019. p. 615-679. DOI: 10.1007/978-981-13-6883-7_24

[5] Food and Agriculture Organization of the United Nations (FAO). Data domain: production: crops. FAO, Rome. Available from: http://www.fao.org/faostat/en/#data/QC. [Accessed: 2018-07-18]

[6] Timsina J, Wolf J, Guilpart N, Van Bussel LG, Grassini P, Van Wart J, Hossain A, Rashid H, Islam S, Van Ittersum MK. Can Bangladesh produce enough cereals to meet future demand?. Agricultural Systems. 2018; 163:36-44. DOI: 10.1016/j.agsy.2016.11.003

[7] Mottaleb KA, Rahut DB, Kruseman G, Erenstein O. Wheat production and consumption dynamics in an Asian rice economy: The Bangladesh case. The European Journal of Development Research. 2018; 30:252-275.

[8] Mottaleb KA, Rahut DB, Kruseman G, Erenstein O. Changing food consumption of households in developing countries: a Bangladesh case. Journal of International Food & Agribusiness Marketing. 2018; 30:156-174. DOI: 10.1080/08974438.2017.1402727

[9] Mottaleb KA, Rahut DB, Kruseman G, Erenstein O. Evolving food consumption patterns of rural and urban households in developing countries. British Food Journal. 2018; 20:392-408. DOI: 10.1108/BFJ-12-2016-0620

[10] Bangladesh Bureau of Statistics (BBS). Statistical year book of Bangladesh. Dhaka: Statistics Division, Ministry of Planning, Government of Peoples Republic of Bangladesh; 2018.

[11] United States Department of Agriculture (USDA). FAS Home/Market and trade data/PSD online. USDA, Foreign Agricultural Service, Washington, DC; 2018. Available from: https://apps.fas.usda.gov/psdonline/app/index.html#/app/advQuery. [Accessed: 2018-09-28]

[12] Barma NCD. An overview on variety development program of WRC. Presented in WRC Internal research review and planning workshop 2017-18, held on 15 July 2018 at BARI seminar room, BARI, Joydebpur, Dhaka.

[13] Intergovernmental Panel on Climate Change (IPCC). Climate change 2007. Synthesis report. In: Pachauri RK, Reisinger AJ editors. Contribution of working groups I, II and III to the fourth assessment report of the intergovernmental panel on climate change. Geneva, 2007. p. 104. Available from: https://www.ipcc.ch/publications_and_data/ar4/wg2/en/ch10s10-4-1.html. [Accessed: 2018-08-15]

[14] Daryanto S, Wang L, Jacinthe PA. Global synthesis of drought effects on maize and wheat production. PloS one. 2016; 11:e0156362. DOI: 10.1371/journal.pone.0156362

[15] Zhang J, Zhang S, Cheng M, Jiang H, Zhang X, Peng C, Lu X, Zhang M, Jin J. Effect of drought on agronomic traits of rice and wheat: a meta-analysis. International Journal of Environmental Research And Public Health. 2018; 15:839. DOI: 10.3390/ijerph15050839

[16] Sarto MVW, Sarto JRW, Rampim L, Rosset JS, Bassegio D, da Costa PF, Inagaki AM. Wheat phenology and yield under drought: a review. Australian Journal of Crop Science. 2017; 11:941. DOI: 10.21475/ajcs17.11.08.pne351

[17] Shahid S, Wang XJ, Harun SB, Shamsudin SB, Ismail T, Minhans A. Climate variability and changes in the major cities of Bangladesh: observations, possible impacts and adaptation. Regional Environmental Change. 2016; 16:459-471. DOI: 10.1007/s10113-015-0757-6

[18] Khan N, Shahid S, Ismail T, Ahmed K, Nawaz N. Trends in heat wave related indices in Pakistan. Stochastic Environmental Research and Risk Assessment. 2019; 33:287-302. DOI: 10.1007/s00477-018-1605-2

[19] Khan N, Shahid S, bin Ismail T, Wang XJ. Spatial distribution of unidirectional trends in temperature and temperature extremes in Pakistan. Theoretical and Applied Climatology. 2019; 136:899-913. DOI: 10.1007/s0070 4-018-2520-7.

[20] Ahmed HGMD, Sajjad M, Li M, Azmat MA, Rizwan M, Maqsood RH, Khan SH. Selection criteria for drought-tolerant bread wheat genotypes at seedling stage. Sustainability. 2019; 11:2584. DOI: 10.3390/su11092584

[21] Shiru MS, Shahid S, Chung ES, Alias N. Changing characteristics of meteorological droughts in Nigeria during 1901-2010. Atmospheric Research. 2019; 223:60-73. DOI: 10.1016/j.atmosres.2019.03.010

[22] Hossain A, Teixeira da Silva JA. Wheat production in Bangladesh: its future in the light of global warming. AoB Plants. 2013; 5. DOI: 10.1093/aobpla/pls042

[23] Abhinandan K, Skori L, Stanic M, Hickerson N, Jamshed M, Samuel MA. Abiotic stress signaling in wheat–an inclusive overview of hormonal interactions during abiotic stress responses in wheat. Frontiers in Plant Science. 2018; 9:734. DOI: 10.3389/fpls.2018.00734

[24] Alam K. Farmers' adaptation to water scarcity in drought-prone environments: A case study of Rajshahi District, Bangladesh. Agricultural Water Management. 2015; 148:196-206. DOI: 10.1016/j.agwat.2014.10.011

[25] Banglapedia, National Encyclopedia of Bangladesh. Available from: http://en.banglapedia.org/ index.php?title=Drought [Accessed: 2020-10-10].

[26] Huq S. Lessons of climate change, stories of solutions: Bangladesh: adaptation. Bulletin of the Atomic Scientists. 2011; 67:56-59. DOI: 10.1177/0096340210393925

[27] Shahid S. Rainfall variability and the trends of wet and dry periods in Bangladesh. International Journal of Climatology. 2010; 30:2299-2313. DOI: 10.1002/joc.2053

[28] Selvaraju R, Baas S. Climate variability and change: adaptation to drought in Bangladesh: A resource book and training guide. Food & Agriculture Organization; 2007.

[29] National Drought Mitigation Center (NDMC). What is drought?

Understanding and defining drought. Available from: http://www.drought.unl.edu/whatis/concept.htm [Accessed: 2017-09-25]

[30] Shahid S. Spatial and temporal characteristics of droughts in the western part of Bangladesh. Hydrological Processes: An International Journal. 2008; 22:2235-2247. DOI: 10.1002/hyp.6820

[31] Shahid S, Behrawan H. Drought risk assessment in the western part of Bangladesh. Natural hazards. 2008; 46:391-413. DOI 10.1007/s11069-007-9191-5

[32] Shahid S, Hazarika MK. Groundwater drought in the northwestern districts of Bangladesh. Water Resources Management. 2010; 24:1989-2006. DOI: 10.1007/s11269-009-9534-y

[33] Dey NC, Alam MS, Sajjan AK, Bhuiyan MA, Ghose L, Ibaraki Y, Karim F. Assessing environmental and health impact of drought in the Northwest Bangladesh. Journal of Environmental Science and Natural Resources. 2011; 4:89-97. DOI: 10.3329/jesnr.v4i2.10141

[34] Habiba U, Shaw R, Takeuchi Y. Farmer's perception and adaptation practices to cope with drought: Perspectives from Northwestern Bangladesh. International Journal of Disaster Risk Reduction. 2012; 1:72-84. DOI: 10.1016/j.ijdrr.2012.05.004

[35] Habiba U, Shaw R, Takeuchi Y. Drought risk reduction through a socio-economic, institutional and physical approach in the northwestern region of Bangladesh. Environmental Hazards. 2011; 10:121-138. DOI: 10.1080/17477891.2011.582311

[36] Khatun M. Climate Change and Migration in Bangladesh: Golden Bengal to Land of Disasters. Bangladesh e-journal of Sociology. 2013; 10.

[37] Shafie H, Halder SR, Rashid AK, Lisa KS, Mita HA. Endowed wisdom: knowledge of nature and coping with disasters in Bangladesh. Dhaka: Center for Disaster Preparedness and Management. 2009.

[38] Ramamasy S, Baas S. Climate variability and change: adaptation to drought in Bangladesh. A resource book and training guide. FAO: Rome, Italy. 2007; p. 68.

[39] Alamgir M, Shahid S, Hazarika MK, Nashrrullah S, Harun SB, Shamsudin S. Analysis of meteorological drought pattern during different climatic and cropping seasons in Bangladesh. JAWRA Journal of the American Water Resources Association. 2015; 51:794-806. DOI: 10.1111/jawr.12276

[40] Rahman MR, Lateh H. Meteorological drought in Bangladesh: assessing, analysing and hazard mapping using SPI, GIS and monthly rainfall data. Environmental Earth Sciences. 2016; 75:1026. DOI: 10.1007/s12665-016-5829-5

[41] Nury AH, Hasan K, Dustegir M, Alam MJ. Drought assessment using standardized precipitation evaporation index and its association with southern oscillation index in the Northwestern Bangladesh. International Journal of Water. 2017; 11:132-158. DOI: 10.1504/IJW.2017.083766

[42] Miah MG, Abdullah HM, Jeong C. Exploring standardized precipitation evapotranspiration index for drought assessment in Bangladesh. Environmental Monitoring and Assessment. 2017; 189:547. DOI: 10.1007/s10661-017-6235-5

[43] Kamruzzaman M, Rahman AS, Ahmed MS, Kabir ME, Mazumder QH, Rahman MS, Jahan CS. Spatio-temporal analysis of climatic variables in the western part of Bangladesh. Environment, Development and

Sustainability. 2018; 20:89-108. DOI: 10.1007/s10668-016-9872-x

[44] Mondol M, Haque A, Ara I, Das SC. Meteorological drought index mapping in Bangladesh using standardized precipitation index during 1981-2010. Advances in Meteorology. 2017; 2017. DOI: 10.1155/2017/4642060

[45] Mohsenipour M, Shahid S, Chung ES, Wang XJ. Changing pattern of droughts during cropping seasons of Bangladesh. Water Resources Management. 2018; 32:1555-1568. DOI: 10.1007/s11269-017-1890-4

[46] Planning Commission, GOB and UNDP Bangladesh. Policy study on climate change on poverty and economic growth and the options of coping with adverse impact of climate change in Bangladesh. Support to Monitoring PRs and MDGs in Bangladesh: Dhaka. 2009.

[47] Mortuza MR, Moges E, Demissie Y, Li HY. Historical and future drought in Bangladesh using copula-based bivariate regional frequency analysis. Theoretical and Applied Climatology. 2019; 135:855-871. DOI: 10.1007/s00704-018-2407-7

[48] Timsina J, Connor DJ. Productivity and management of rice–wheat cropping systems: issues and challenges. Field Crops Research. 2001; 69:93-132. DOI: 10.1016/S0378-4290(00)00143-X

[49] Ahammed SJ, Homsi R, Khan N, Shahid S, Shiru MS, Mohsenipour M, Ahmed K, Nawaz N, Alias NE, Yuzir A. Assessment of changing pattern of crop water stress in Bangladesh. Environment, Development and Sustainability. 2019; 1-9. DOI: 10.1007/s10668-019-00400-w

[50] Ahmed AU. Bangladesh climate change impacts and vulnerability. A synthesis; climate change cell; department of environment, comprehensive disaster management programme: Dhaka, Bangladesh. 2006.

[51] Timsina J, Jat ML, Majumdar K. Rice-maize systems of South Asia: current status, future prospects and research priorities for nutrient management. Plant and Soil. 2010; 335:65-82. DOI: 10.1007/s11104-010-0418-y

[52] Timsina J, Buresh RJ, Dobermann A, Dixon J. Rice-maize systems in Asia: current situation and potential. IRRI, Los Banos, Philippines. 2011. p. 235.

[53] Kabir MJ, Islam MM. Study on agronomically and economically dominant cropping patterns in some selected areas of Barisal district. Bangladesh Journal of Agricultural Research. 2012; 37:55-65. DOI: 10.3329/bjar.v37i1.11177

[54] Anjum SA, Xie XY, Wang LC, Saleem MF, Man C, Lei W. Morphological, physiological and biochemical responses of plants to drought stress. African Journal of Agricultural Research. 2011; 6:2026-2032. DOI: 10.5897/AJAR10.027

[55] Liu H, Sultan MARF, Liu XL, Zhang J, Yu F, Zhao HX. Physiological and comparative proteomic analysis reveals different drought responses in roots and leaves of drought-tolerant wild wheat (*Triticum boeoticum*). PLoS One. 2015; 10:e0121852. DOI:10.1371/journal.pone.0121852

[56] Suzuki N, Rivero RM, Shulaev V, Blumwald E, Mittler R. Abiotic and biotic stress combinations. New Phytologist. 2014; 203:32-43. DOI: 10.1111/nph.12797

[57] Tardieu F, Granier C, Muller B. Water deficit and growth. Co-ordinating processes without an orchestrator?. Current Opinion in Plant Biology. 2011; 14:283-289. DOI: 10.1016/j.pbi.2011.02.002

[58] Vadez V, Kholova J, Zaman-Allah M, Belko N. Water: the most important

'molecular' component of water stress tolerance research. Functional Plant Biology. 2013; 40:1310-1322. DOI: 10.1071/FP13149

[59] Nezhadahmadi A, Prodhan ZH, Faruq G. Drought tolerance in wheat. The Scientific World Journal. 2013; 2013:1-12. DOI: 10.1155/2013/610721

[60] Akram M. Growth and yield components of wheat under water stress of different growth stages. Bangladesh Journal of Agricultural Research. 2011; 36:455-468. DOI: 10.3329/bjar.v36i3.9264

[61] Saeidi M, Abdoli M. Effect of drought stress during grain filling on yield and its components, gas exchange variables, and some physiological traits of wheat cultivars. Journal of Agricultural Science and Technology. 2015; 17:885-898.

[62] Wang X, Vignjevic M, Liu F, Jacobsen S, Jiang D, Wollenweber B. Drought priming at vegetative growth stages improves tolerance to drought and heat stresses occurring during grain filling in spring wheat. Plant Growth Regulation. 2015; 75:677-687. DOI: 10.1007/s10725-014-9969-x

[63] Shamsi K, Kobraee S. Bread wheat production under drought stress conditions. Annals of Biological Research. 2011; 2:352-358.

[64] Ihsan MZ, El-Nakhlawy FS, Ismail SM, Fahad S. Wheat phenological development and growth studies as affected by drought and late season high temperature stress under arid environment. Frontiers in Plant Science. 2016; 7:795. DOI: 10.3389/fpls.2016.00795

[65] Chen D, Wang S, Cao B, Cao D, Leng G, Li H, Yin L, Shan L, Deng X. Genotypic variation in growth and physiological response to drought stress and re-watering reveals the critical role of recovery in drought adaptation in maize seedlings. Frontiers in Plant Science. 2016; 6:1-15. DOI: 10.3389/fpls.2015.01241

[66] Abid M, Tian Z, Ata-Ul-Karim ST, Wang F, Liu Y, Zahoor R, Jiang D, Dai T. Adaptation to and recovery from drought stress at vegetative stages in wheat (*Triticum aestivum*) cultivars. Functional Plant Biology. 2016; 43:1159-1169. DOI: 10.1071/FP16150

[67] Izanloo A, Condon AG, Langridge P, Tester M, Schnurbusch T. Different mechanisms of adaptation to cyclic water stress in two South Australian bread wheat cultivars. Journal of Experimental Botany. 2008; 59:3327-3346. DOI:10.1093/jxb/ern199

[68] Khanna-Chopra R, Selote DS. Acclimation to drought stress generates oxidative stress tolerance in drought-resistant than-susceptible wheat cultivar under field conditions. Environmental and Experimental Botany. 2007; 60:276-283. DOI: 10.1016/j.envexpbot.2006.11.004

[69] Maxwell K, Johnson GN. Chlorophyll fluorescence – a practical guide. Journal of Experimental Botany. 2000; 51:659-668. DOI: 10.1093/jexbot/51.345.659

[70] Galmés J, Flexas J, Savé R, Medrano H. Water relations and stomatal characteristics of Mediterranean plants with different growth forms and leaf habits: responses to water stress and recovery. Plant and Soil. 2007; 290:139-155. DOI: 10.1007/s11104-006-9148-6

[71] Ali A, Syed AAW, Khaliq T, Asif M, Aziz M, Mubeen M. Effects of nitrogen on growth and yield components of wheat (report). Biological Sciences. 2011; 3:1004-1005.

[72] Kantar M, Lucas SJ, Budak H. miRNA expression patterns of *Triticum*

dicoccoides in response to shock drought stress. Planta. 2011; 233:471-484. DOI: 10.1007/s00425-010-1309-4

[73] Ludlow MM. Strategies of response to water stress. In: Kreeb KH, Richter H, Hinckley TM, editors. Structural and functional responses to environmental stresses: water shortage. SPB Academic, The Hague; 1989. p. 269-281.

[74] Blum A. Plant breeding for water-limited environments. Springer Science & Business Media, New York. 2010.

[75] Comas L, Becker S, Cruz VM, Byrne PF, Dierig DA. Root traits contributing to plant productivity under drought. Frontiers in Plant Science. 2013; 4:442. DOI: 10.3389/fpls.2013.00442

[76] Henry A, Cal AJ, Batoto TC, Torres RO, Serraj R. Root attributes affecting water uptake of rice (*Oryza sativa*) under drought. Journal of Experimental Botany. 2012; 63:4751-4763. DOI: 10.1093/jxb/ers150

[77] Manschadi AM, Christopher J, deVoil P, Hammer GL. The role of root architectural traits in adaptation of wheat to water-limited environments. Functional Plant Biology. 2006; 33:823-837. DOI: 10.1071/FP06055

[78] Lilley JM, Kirkegaard JA. Benefits of increased soil exploration by wheat roots. Field Crops Research. 2011; 122:118-130. DOI: 10.1016/j.fcr.2011.03.010

[79] Ali ML, Luetchens J, Singh A, Shaver TM, Kruger GR, Lorenz A. Greenhouse screening of maize genotypes for deep root mass and related root traits and their association with grain yield under water-deficit conditions in the field. Euphytica. 2016; 207:79-94. DOI: 10.1007/s10681-015-1533-x

[80] Ayalew H, Ma X, Yan G. Screening wheat (*Triticum* spp.) genotypes for root length under contrasting water regimes: potential sources of variability for drought resistance breeding. Journal of Agronomy and Crop Science. 2015; 201:189-194. DOI: 10.1111/jac.12116

[81] Li R, Zeng Y, Xu J, Wang Q, Wu F, Cao M, Lan H, Liu Y, Lu Y. Genetic variation for maize root architecture in response to drought stress at the seedling stage. Breeding Science. 2015; 65:298-307. DOI: 10.1270/jsbbs.65.298

[82] Siddique KHM, Belford RK, Tennant D. Root: shoot ratios of old and modern, tall and semi-dwarf wheats in a Mediterranean environment. Plant and Soil. 1990; 121:89-98. DOI: 10.1007/BF00013101

[83] Ehdaie B, Layne AP, Waines JG. Root system plasticity to drought influences grain yield in bread wheat. Euphytica. 2012; 186:219-232. DOI: 10.1007/s10681-011-0585-9

[84] Richards RA, Rebetzke GJ, Condon AG, Van Herwaarden AF. Breeding opportunities for increasing the efficiency of water use and crop yield in temperate cereals. Crop Science. 2002; 42:111-121. DOI: 10.2135/cropsci2002.1110

[85] Rampino P, Pataleo S, Gerardi C, Mita G, Perrotta C. Drought stress response in wheat: physiological and molecular analysis of resistant and sensitive genotypes. Plant, Cell & Environment. 2006; 29:2143-2152. DOI: 10.1111/j.1365-3040.2006.01588.x

[86] Faisal SU, Mujtaba SM, Khan MA, Mahboob WA. Morpho-physiological assessment of wheat (*Triticum aestivum* L.) genotypes for drought stress tolerance at seedling stage. Pakistan Journal of Botany. 2017; 49:445-452.

[87] Ahmed HG, Zeng Y, Yang X, Anwaar HA, Mansha MZ, Hanif CM, Ikram K, Ullah A, Alghanem SM. Conferring drought-tolerant wheat genotypes through morpho-physiological

and chlorophyll indices at seedling stage. Saudi Journal of Biological Sciences. 2020; 27:2116-2023. DOI: 10.1016/j.sjbs.2020.06.019

[88] Zhu M, Shabala S, Shabala L, Fan Y, Zhou MX. Evaluating predictive values of various physiological indices for salinity stress tolerance in wheat. Journal of Agronomy and Crop Science. 2016; 202:115-124. DOI: 10.1111/jac.12122

[89] Khadka K, Earl HJ, Raizada MN, Navabi A. A Physio-morphological trait-based approach for breeding drought tolerant wheat. Frontiers in Plant Science. 2020; 11:715. DOI: 10.3389/fpls.2020.00715

[90] Khan MM, Khan MSI, Mondal RS, Rashid MH, Faruq G, Rahman MM, Barma NCD. Performance of some selected wheat genotypes in southern Bangladesh. WRC Internal Research Review Report 2017-18 (Crop Management). Wheat Research Center, Bangladesh Agricultural Research Institute, Nashipur, Dinajpur-5200, Bangladesh. 2018.

[91] Cattivelli L, Rizza F, Badeck FW, Mazzucotelli E, Mastrangelo AM, Francia E, Marè C, Tondelli A, Stanca AM. Drought tolerance improvement in crop plants: an integrated view from breeding to genomics. Field Crops Research. 2008; 105:1-14. DOI: 10.1016/j.fcr.2007.07.004

[92] Sserumaga JP, Beyene Y, Pillay K, Kullaya A, Oikeh SO, Mugo S, Machida L, Ngolinda I, Asea G, Ringo J, Otim M. Grain-yield stability among tropical maize hybrids derived from doubled-haploid inbred lines under random drought stress and optimum moisture conditions. Crop and Pasture Science. 2018; 69:691-702. DOI: 10.1071/CP17348

[93] Reynolds MP, Pellegrineschi A, Skovmand B. Sink-limitation to yield and biomass: a summary of some investigations in spring wheat. Annals of Applied Biology. 2005; 146:39-49. DOI: 10.1111/j.1744-7348.2005.03100.x

[94] Reynolds MP, Trethowan RM. Physiological interventions in breeding for adaptation to abiotic stress. In: Spiertz JHJ, Struik PC, van Laar HH, editors. Scale and complexity in plant systems research: gene-plant-crop relations. Cham: Springer; 2007. p. 129-146. DOI: 10.1002/anie.199315241

[95] Dolferus R, Thavamanikumar S, Sangma H, Kleven S, Wallace X, Forrest K, Rebetzke G, Hayden M, Borg L, Smith A, Cullis B. Determining the genetic architecture of reproductive stage drought tolerance in wheat using a correlated trait and correlated marker effect model. G3: Genes, Genomes, Genetics. 2019; 9:473-489. DOI: 10.1534/g3.118.200835

[96] Wieczorek A. Use of Biotechnology in Agriculture-Benefits and Risks. University of Hawaii, Biotechnology, BIO-3, Honolulu, Hawaii, USA, 2003.

[97] Huang B. Gametoclonal variation in crop improvement. In: Jain SM, Sopory SK, Veilleux RE, editors. In vitro haploid production in higher plants. vol 2. Kluwer, Dordrecht; 1996. p. 73-91.

[98] Raina SK. Doubled haploid breeding in cereals. In: Janick J, editor. Plant breeding reviews. vol 15. Wiley, New York; 1997. p. 141-186.

[99] Ma H, Busch RH, Riera-Lizarazu O, Rines HW, Dill-Macky R. Agronomic performance of lines derived from anther culture, maize pollination and single-seed descent in a spring wheat cross. Theoretical and Applied Genetics. 1999; 99:432-436.

[100] Sobieh SS. Induction of short culm mutants for bread wheat by using gamma rays. Arab Journal of Nuclear Sciences and Applications. 2002; 35:318-328.

[101] Al-Naggar AM, Ragab AE, Youssef SS, Al-Bakry RI. New genetic variation in drought tolerance induced via irradiation and hybridization of Egyptian cultivars of bread wheat. Egyptian Journal of Plant Breeding. 2004; 8:353-370.

[102] Al-Naggar AM, Atta MM, Shaheen AM, Al-Azab KF. Gamma rays and EMS induced drought tolerant mutants in bread wheat. Egyptian Journal of Plant Breeding. 2007; 11:135-165.

[103] FAO/IAEA. Mutant variety database. Cereals and legumes. FAO/IAEA, Vienna. 2012. Available from: http://mvgs.iaea.org

[104] Sakin MA, Gokmen S, Yildirim A. Investigation of mutants induced in durum wheat (*Triticum durum* Desf.) for yield and some agronomic and quality traits. Asian Journal of Plant Sciences. 2005; 4:279-283.

[105] Ahmed HG, Khan AS, Khan SH, Kashif M. Genome wide allelic pattern and genetic diversity of spring wheat genotypes through SSR markers. International Journal of Agriculture & Biology. 2017; 19:1559-1565. DOI: 10.17957/IJAB/15.0463

[106] Ashraf M. Inducing drought tolerance in plants: recent advances. Biotechnology advances. 2010; 28:169-183. DOI: 10.1016/j.biotechadv.2009.11.005

[107] Powell W, Morgante M, Andre C, Hanafey M, Vogel J, Tingey S, Rafalski A. The comparison of RFLP, RAPD, AFLP and SSR (microsatellite) markers for germplasm analysis. Molecular Breeding. 1996; 2:225-238. DOI: 10.1007/BF00564200

[108] Russell JR, Fuller JD, Macaulay M, Hatz BG, Jahoor A, Powell W, Waugh R. Direct comparison of levels of genetic variation among barley accessions detected by RFLPs, AFLPs, SSRs and RAPDs. Theoretical and Applied Genetics. 1997; 95:714-722.

[109] Davila JA, Loarce Y, Ferrer E. Molecular characterization and genetic mapping of random amplified microsatellite polymorphism in barley. Theoretical and Applied Genetics. 1999; 98:265-273.

[110] Ibrahim SE, Schubert A, Pillen K, Léon J. QTL analysis of drought tolerance for seedling root morphological traits in an advanced backcross population of spring wheat. International Journal of Agricultural Science. 2012; 2:619-629.

[111] Ahmad MQ, Khan SH, Khan AS, Kazi AM, Basra SMA. Identification of QTLs for drought tolerance traits on wheat chromosome 2A using association mapping. International Journal of Agriculture and Biology. 2014; 16: 862-870.

[112] Kumar U, Joshi AK, Kumari M, Paliwal R, Kumar S, Röder MS. Identification of QTLs for stay green trait in wheat (*Triticum aestivum* L.) in the 'Chirya 3'×'Sonalika'population. Euphytica. 2010; 174:437-445. DOI: 10.1007/s10681-010-0155-6

[113] Sharma RK. Does low yield heterosis limit commercial hybrids in wheat?. African Journal of Agricultural Research. 2013; 8: 6663-6669. DOI: 10.5897/AJAR2013.8108

[114] Gosal SS, Wani SH, Kang MS. Biotechnology and drought tolerance. Journal of Crop Improvement. 2009; 23:19-54. DOI: 10.1080/15427520802418251

[115] Kereša S, Barić M, Horvat M, Habuš Jerčić I. Drought tolerance mechanisms in plants and their genetic base in wheat. Sjemenarstvo. 2008; 25:35-45.

[116] Mahpara S, Hussain ST, Farooq J. Drought tolerance studies in wheat (*Triticum aestivum* L.). Cercetari Agronomice in Moldova. 2015; 47:133-140. DOI: 10.1515/cerce-2015-0011

[117] Wang K, Riaz B, Ye X. Wheat genome editing expedited by efficient transformation techniques: progress and perspectives. The Crop Journal. 2018; 6:22-31. DOI: 10.1016/j.cj.2017.09.009

[118] Mir RR, Zaman-Allah M, Sreenivasulu N, Trethowan R, Varshney RK. Integrated genomics, physiology and breeding approaches for improving drought tolerance in crops. Theoretical and Applied Genetics. 2012; 125:625-645. DOI: 10.1007/s00122-012-1904-9

[119] Kosova K, Vitamvas P, Urban MO, Kholova J, Prášil IT. Breeding for enhanced drought resistance in barley and wheat-drought-associated traits, genetic resources and their potential utilization in breeding programmes. Czech Journal of Genetics and Plant Breeding. 2014; 50:247-261. DOI: 10.17221/118/2014-cjgpb

[120] Choudhary AK, Sultana R, Vales MI, Saxena KB, Kumar RR, Ratnakumar P. Integrated physiological and molecular approaches to improvement of abiotic stress tolerance in two pulse crops of the semi-arid tropics. The Crop Journal. 2018; 6:99-114. DOI: 10.1016/j.cj.2017.11.002

Chapter 9

Energy Use Efficiency in Irrigated and Rainfed Wheat in Pakistan

Muhammad Imran and Orhan Özçatalbaş

Abstract

Wheat is an important staple food in Pakistan and is grown in both irrigated and rainfed production systems. To meet increased demand, farmers have increased the use of input energy in wheat production. The intensive use of energy has many consequences for energy security and environmental sustainability. In this chapter, we have analyzed the energy use efficiency of wheat crop grown in two different production systems using data collected from wheat farmers of Punjab province of Pakistan through face-to-face interviews. Energy input–output analysis revealed that 49,079 MJ/ha input energy is used in irrigated wheat and 31,421 MJ/ha in rainfed wheat. The main difference between both production systems is because of irrigation water. Fertilizer has the highest share in total energy consumption followed by diesel fuel. Energy consumed per kilogram of wheat produced is less in rainfed wheat compared to irrigated. Similarly, energy efficiency values of rainfed wheat are better than irrigated wheat. Results of data envelopment analysis reveal that 38% of wheat farmers in rainfed systems and 62% in the irrigated system are using energy efficiently. The substantial difference between the energy use of inefficient and efficient indicates that there's a significant potential to improve energy use efficiency in both systems.

Keywords: energy use efficiency, input–output analysis, DEA, wheat, Pakistan

1. Introduction

Population growth and increased demand for food have led humanity to look for new ways to increase food production. Energy, which is an essential input in agriculture, has been considered as a feasible option to increase food productivity and enhance food security. As a result, agriculture has become energy-intensive to meet increased food and biofuel demand [1].

After the green revolution, the introduction of high yield varieties and intensive crop management practices has increased the use of energy manifolds in both developing and developed countries [2, 3].

It is anticipated that energy input for crop production will increase further mainly due to population and economic growth, climate change, degrading quality of soils, and shortage of labor [4, 5]. On the other hand, intensive use of energy in crop production is posing many threats to agriculture sustainability, human health, and sustainability of the environment. Sometimes to get maximum returns farmers make overuse of energy inputs. This has led to increased energy used in crop production at a faster rate compared to other sectors. Escape of traditional practices in agriculture, technological advancements in Agri-machinery, and increased application rate of fertilizer is also responsible for increased use of energy in crop production. It is also ascribed to the introduction of high yielding varieties, and excessive use of biocides and chemical fertilizer. In addition to this diesel fuel consumption has also increased due to farm mechanization and pumping of underground water. Finally, scarcity of cultivable lands and irrigation water increased the human population, and the desire for improved living standards has also contributed to the intensive use of energy in agriculture. Both agriculture and the environment are dependent on each other and the efficient use of energy is a basic requirement for sustainable agriculture [6, 7]. Sustainable development of agriculture is dependent on high energy use efficiency with low energy use in crop production. Thus, increasing energy use efficiency in crop production is important for food security and environmental sustainability. Keeping in view the multiple interactions of agriculture with the environment, analysis of the consumption of energy (both operational and embodied) in the agriculture system is urgently needed to fight both environmental issues stemming from agriculture and climate change impacts on agriculture.

1.1 Environmental implications of input energy use in agriculture

Agriculture contributes 24% of global Greenhouse gases emission, and agricultural activities are considered a significant source of pollution [8, 9]. It is estimated that GHG emission from agriculture has doubled in the last 50 years, they could increase by another 30% by 2050 [10]. Increasing use of energy inputs in agriculture is associated with numerous environmental problems such as loss of biodiversity, pollution of the aquatic environment by chemical fertilizers and pesticides, and high consumption of non-renewable energy resources. Among all other energy inputs used in crop production, diesel fuel and fertilizers have the highest share of energy consumption [11, 12]. Studies have found that fertilizer and pesticides are among the most substantial secondary sources of CO2 emissions [8]. According to an intergovernmental panel on climate change [13]. Direct and indirect consumption of fossil fuels for crop production leads to the emission of carbon dioxide (CO2), nitrous oxide (NO2), and methane (CH4). Climate Change resulting from greenhouse gasses is the most important environmental challenges in today's world [13]. A significant portion of these greenhouse gases is produced by agriculture. About 10–12% of all anthropogenic GHG emissions are contributed by agricultural greenhouse gasses emission [14].

The major use of commercial energy in agriculture is during the production and operation of agricultural machinery. Most of the agricultural operations like, land preparation, irrigation, fertilization, spraying, and harvesting are performed using fossil fuels. The combustion of fossil fuels in agricultural machinery releases CO2 into the atmosphere.

Excessive or over-use of fertilizers leads to loss of nutrient elements, which are main contributors to non-point source pollution from agriculture, degradation of water and soil quality, decrease in the quality of agricultural products, and increase in air emissions. Due to losses incurred by pest attacks, the use of pesticides is increasing at a higher rate. There is a 4.4% average annual growth in the use of agrochemicals worldwide [15]. This increased use of pesticides is causing air, water, and soil pollution. The increasing use of pesticides in agriculture is becoming the main environmental hazard and a major contributor to agriculture pollution. Additionally, agriculture is thought to be the major contributor of N2O by indirect and direct sources [16]. The food production system is under increasing pressure due to consistent population growth and climate change; by an increase in demand for food security while protecting the natural resources by minimizing the environmental footprints [17].

Both sustainable environment and sustainable agriculture are dependent on each other. Environmental factors have a significant contribution to agriculture; agriculture, as compared to other sectors, is more dependent on the natural environment. Agriculture is the source of food and fiber for the human being and vital for human existence; as a result, sustainable agriculture development is not just related to economic development but also human survival. Therefore, efficient use of energy is one of the conditions for sustainable agriculture [18].

1.2 Energy efficiency in agriculture

Efficient use of energy inputs helps to increase production and productivity, profitability and competitiveness of agriculture, and sustainable rural living. Higher energy use efficiency will promote sustainable agriculture by minimizing environmental problems and preventing the destruction of natural resources. The use of renewable energy sources and increase in efficiency of energy can also make a significant contribution in achieving sustainable energy development goals [19]. Currently, the world is focused to develop a production system that maintains high levels of output while minimizing the input of fossil energy and as a result, helps to reduce greenhouse gas emissions. To combat global warming, reducing emissions of greenhouse gases by minimizing the direct and indirect use of fossil fuels for crop production is a vital strategy. Energy efficiency is an essential element for achieving sustainable agricultural development. This is also important for increasing economic returns, preserving fossil fuel reserves, and sustainable agricultural production. Therefore, environmental impact assessments, energy analysis, and GHG emission assessments are important components.

2. Wheat production in Pakistan

Wheat (*Triticum aestivum* L.) is an important winter crop in Pakistan. Wheat significantly contributes to the livelihood and food security of the population in Pakistan, as well as at the global and regional levels. It meets about 1/5th of the daily calorie and protein requirement of human beings [20] and it constitutes 65% of staple food consumption in Pakistan. It contributes 1.7% to the national GDP of Pakistan and 8,7% to agriculture value addition. Wheat was cultivated on 8,25 Million hectares in 2019–2020 and the area under wheat has slightly decreased in the past five years. Over the years, wheat yield per acre has been stagnant or little change has been seen due to declined under-ground water table, soil degradation, environmental pollution, etc. delayed sowings, low germination rate, insect-pest infestation, and low crop stand has lowered the production efficiency of wheat. A further decline in wheat yield in recent years can be attributed to locust attacks. Keeping in view increasing population and government policies (increased support price from 1400/40 kg to 1650/40 kg before the wheat season in 2020), it is projected that farmer will divert their resource towards wheat to get maximum output from a limited quantity of arable land. The limited supply of labor on one hand and incentives for higher productivity on other hand will lead to increased use of energy in wheat production. In Pakistan, winter wheat is grown both irrigated and drylands. During winter availability of canal water is almost negligible and irrigated wheat is irrigated with groundwater. However, sustainability productivity of wheat crop is under threat due to over-exploitation of underground water. Moreover, a substantial amount of diesel fuel is used to pump water from underground, leading to significant consumption of diesel fuel energy in wheat production. On the other hand, water is a scarce resource and the water table is depleting rapidly in Pakistan. These both issues are posing a great threat to the environmental sustainability of Pakistan, as Pakistan is among the 10 most climate affected countries in the world. The worsening energy and water issue in Pakistan needs the urgent attention of policymakers.

2.1 Input energy use in wheat production

There's substantial use of energy in wheat production both directly and indirectly. In operations like tillage, planting, and harvesting there's a direct use of energy, while energy is indirectly used in inputs such weedicides, fertilizers, and agriculture machinery (**Figure 1**).

2.1.1 Human labor

Human labor is the most important source of the energy in agriculture, although the introduction of machines has reduced human labor in the industry in the field activities, human labor is still playing its key role. In agricultural activities, human labor is used almost at every step, from manual work on the farm, driving

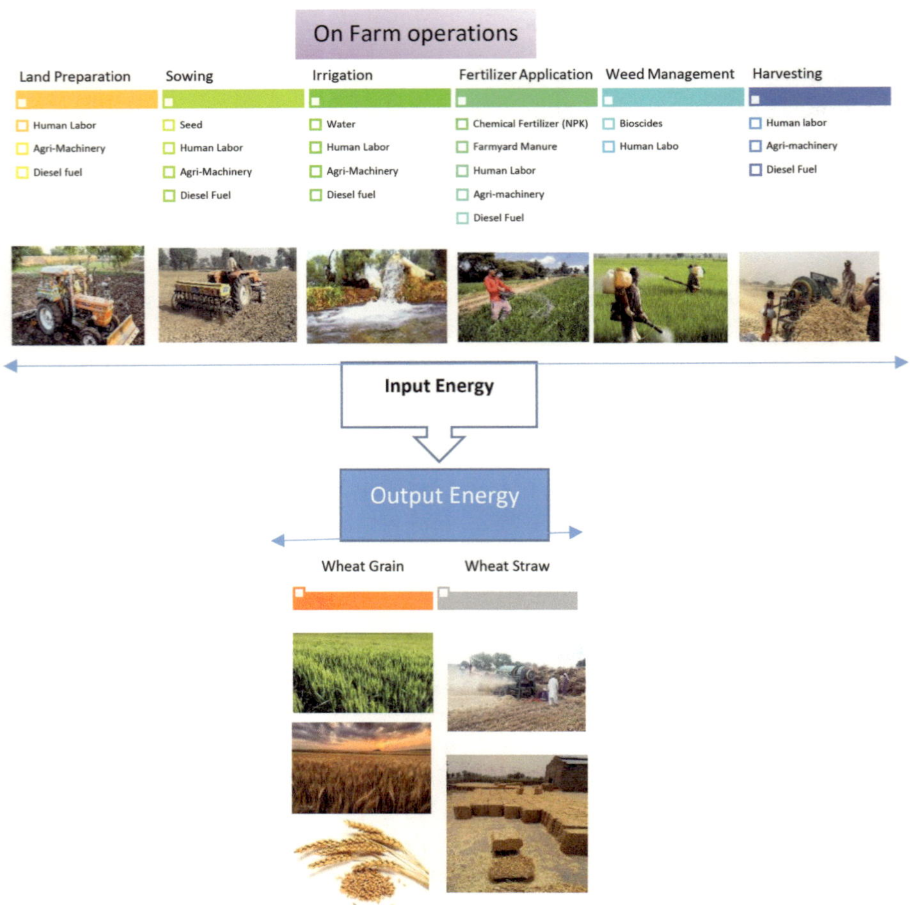

Figure 1.
System boundaries of wheat production system in Pakistan.

Inputs	Mean (S.E)	Min.	Max.	Energy equivalents
Human Labor (hours)	178.45 (6.38)	3.89	391.82	1.96
Seed (kg)	134.19 (0.86)	123.5	148.50	15.7
Diesel fuel (liter)	139.98(3.94)	29.64	397.67	56.31
Irrigation water (m^3)	8483.07 (3887)	0	612,809	1.02
Fertilizer (all)	345.15 (10.52)	0	741	
Nitrogen (kg)	177.68 (7.39)	0	617.50	66.14
Phosphate (kg)	130.7 (4.160	0	370.50	12.44
Potash (kg)	37.36 (5.36)	0	370.50	11.15
Herbicides (kg)	1.60 (0.10)	0	4.94	278
Farmyard manure (kg)	30,982.5 (2668)	0	180,000	0.3

Table 1.
Quantity of inputs used in wheat production in Pakistan and their energy equivalents.

agricultural machinery, maintenance, fertilizer and pesticide application, irrigation, and harvesting to management. In developing countries, human power constitutes 73% of the total energy use on farms [21]. Maybe in the future with full mechanization of farms, the use of human labor will be reduced, but some scientists believe that organic and modern agriculture needs more manual work for weeding and harvesting [22, 23]. There are different estimates for the energy output of human labor on farms. The main physical activities in wheat production are driving a tractor, manual sowing, manual fertilization and spraying, harvesting, and transportation. In this study, human labor work was calculated based on the information provided by the wheat farmers on the number of hours spent in each operation. The energy equivalent of human labor is muscle power used in the field operations of crop production. The energy equivalent of human labor is 1.96 MJ/h determined from literature (**Table 1**). Labor energy consumption can be determined by multiplying total hours of human activity by the energy coefficients of workers. In Pakistan, where still mechanization of the farms is not so common, there is ample use of human labor in the farm operations. On average 178.45 hours of human labor is used in one hectare of wheat production.

2.1.2 Seed

Seed is mostly provided by seed producers and private seed companies; however, some farmers also use seeds from their farms. Wheat is planted either by seed drill or manually by spreading, the amount of seed also varies according to the sowing method. On average, 134.19 kg/ha wheat seed is used in Pakistan. Energy equivalents of the seed are the energy used in the preparation of wheat seed. Energy inputs of seed can be calculated by multiplying the quantity of seed used per hectare with its energy equivalents (8.65 MJ/kg).

2.1.3 Farm machinery

The embedded energy necessary to manufacture machinery for crop production is a tertiary input that typically has a minor impact on the total energy. Farrell et al. [24] reported that machinery accounted for only 1.7% of the total energy associated with corn production. Therefore, energy use in machinery is not included in the estimation of energy used in wheat production.

2.1.4 Fossil fuels

Diesel fuel is the main fuel used in farm machinery and water pump for different crop operations. Consumption of the fuel is dependent on several factors like climate, crop, soil, rolling assistance, and speed. In dry and warm climate use of diesel is more for irrigation than other operations, while in dry farming system diesel is mainly used in tillage and sowing as compared to irrigation. The energy output of diesel fuel was calculated by multiplying liter/ha with fuel equivalent of energy per liter. Energy equivalents of diesel fuel are 44.83 MJ/L. The average diesel fuel use is 39.98 liter/ha in wheat production.

2.1.5 Fertilizer chemical and pesticides

Soil nutrients are the most important obstacle to crop productivity. Fertilizers are used by farmers to increase soil nutrients and resultant growth. Chemical, organic, and biological fertilizers are used in crop production, but just chemical fertilizers are believed to increase the yield more than any other fertilizer. Nitrogen is the main mineral fertilizer being used in crop production. Nitrogen fertilizer is energy-intensive, on the other hand, phosphate and potash do not need high energy. Chemical and chemical fertilizers energy equivalents mean the energy consumption for production, packing, and distribution of the material. On average 177.68 kg per hectare of nitrogen nutrients, 130.17 kg phosphate nutrients, and 37.36 kg potash are used in wheat production in Pakistan. Additionally, 1.60 kg per hectare of herbicides are used in wheat production for weed management.

2.1.6 Water for irrigation

While dry-land wheat is dependent on rains, but irrigated wheat requires irrigation water throughout the production process. On average 8483.07 m^3 of irrigation water is used in one hectare of wheat. The energy equivalents of the water for irrigation input is the indirect energy of irrigation consists of the energy consumed for manufacturing the material for the dams, canals, pipes, pumps, and equipment as well as the energy for constructing the walls and building the on-farm irrigation system. The energy equivalent of the irrigation was estimated to be 0.014 MJ/m^3.

2.2 Energy balances in wheat production

Energy consumption in wheat production includes; labor, embodied energy in seed, chemical and fertilizers, diesel, and water for irrigation. Except water for irrigation all other input energies are same for rainfed (dry land) wheat. There's a wide variation of input energy (**Table 2**), which shows high level of mismanagement in usage of energy resources among some wheat producers. This also indicates that there is great scope for improving energy consumption efficiencies of wheat producers in both farming systems. On average total input energy consumption in irrigated wheat is 49,079.27 MJ ha^{-1} and 31421.59 MJ ha^{-1} for rainfed wheat. The higher use of input energy use in irrigated wheat can be attributed to irrigation energy. Highest share of energy consumption in irrigated wheat is from chemical fertilizer (31.33%), while farmyard manure contributes highest in total input energy consumption in rainfed wheat.

In fertilizers, nitrogen constitutes the highest share, 80.39% and 82.31%, in irrigated and rain-fed wheat, respectively. Highest share of nitrogen in total fertilizer

Energy Inputs	Irrigated		Rain-fed	
	Energy equivalents MJ ha^{-1}	SD*	Energy equivalents MJ ha^{-1}	SD*
Human labor	402.07	166.78	259.45	163.12
Seed	2157.54	193.91	2017.93	157.72
Diesel fuel	9435.13	2697.53	5155.56	1835.76
Water for irrigation	13578.13	7578.43	—	—
Chemicals	627.10	358.56	129.87	324.53
Farmyard manure	7518.00	10767.05	12837.32	12363.56
Nitrogen	13069.26	6998.60	9437.68	6374.82
Phosphate	1702.02	675.63	1474.07	1015.25
Potash	589.68	994.91	109.69	354.96
Yield (output)	50756.79	11715.46	34427.32	20161.36

**Standard Deviation.*

Table 2.
Energy balance in both production systems.

consumption is also recorded in some other countries by [25–27]. Though, nitrogen fertilizer has played key role in enhancing the food production, at the same time excessive use of nitrogen has contributed to soil, water, and air pollution in many parts of the world. Sustainability of crop production is threatened by overuse of inorganic fertilizer which inflicts severely on soil health. The need for nitrogen can be reduced by fertilization management and integrating a legume in crop rotation. In order to reduce demand for inorganic fertilizer in medium term, soil fertility and organic matter contents can be increased by applying composts, chopped residues or other soil amendments. Almost, 55% of the farmers in Punjab (Pakistan) just use inorganic fertilizers, and 30% use combination of both organic and inorganic. Furthermore, farmers use more than recommended dose of fertilizer (Zulfiqar et al. 2017). So, adopting balanced use of fertilizer by wheat producers will reduce the use of nitrogen, as nitrogen has been found to be main difference between conventional and sustainable farming system (Pimentel et al. 2005). So, consumption of nitrogen with organic fertilizer and balanced use of fertilizer will reduce energy consumption in production system and improve its productivity.

Water for irrigation is the second largest consumer of energy in irrigated wheat. Diesel fuel is used for operating machinery in wheat production, it constitutes 19.25% of the total input energy consumption in irrigated and 16.4% in rain-fed. Börjesson and Tufvesson [28] found diesel as the main energy input after fertilizer in wheat, sugar beet, canola and maize. Particularly in irrigated land where diesel is also used for ground water pumping its use is higher (9435.13 MJ ha^{-1}) than rain-fed (1835.76 MJ ha^{-1}). Siddiqi and Wescoat [29] reported that ground water pumping consumes 61% of direct energy in Punjab. Pumping systems are mostly dependent on fossil fuels, almost 91% of the total installed pumps use diesel driven motors.

Furthermore, share of human labor (0.81%) with amount of 402.07 MJ ha-1 in the irrigated farming system is the least in total energy consumption, followed by chemicals and seed. In rain-fed wheat share of chemical (0.4%) in total energy consumption was negligible followed by human labor and seed. The average output energy in irrigated wheat was calculated as 50756.79 MJ ha-1, and 34427.32 MJ ha-1 for rain-fed wheat farming.

2.3 Energy indices

Energy ratio which is a relationship between input and output energy is often used as an index to measure energy efficiency in crop production. Energy ratio can also be used to determine subsistence of the system in isolated societies. If ratio is lower than one, it means system is losing energy and if it is higher than one it means system is earning energy. Energy efficiency for irrigated and rain-fed wheat production is estimated to be 1.03 and 1.09, respectively (**Table 3**). Irrigation can be the reason for difference between two production system, higher energy efficiency for rain-fed and comparatively low for irrigated. This suggests that an efficient irrigation system will improve energy ratio in irrigated wheat. For comparisons between two production system energy efficiency may not be very good approach, because difference in energy efficiency can be due to difference in energy input and yield. Ziaei et al. [30] said that energy productivity is comparatively a better parameter to show the difference between two production systems, as it calculates the ratio of production yield per kg into consumer energy. Estimates of energy productivity shows that, for each unit of input energy (MJ) consumed in wheat, 0.07 and 0.06 yield units are achieved in rain-fed and irrigated wheat production, respectively (**Table 3**). This again shows that, energy is more efficiently being used in rainfed production system. Specific energy was estimated to be 12.70 and 14.49 MJ kg^{-1} for rain-fed and irrigated wheat production (**Table 3**). Lower value of specific energy shows that less amount of energy is used for production of one yield unit, as it is reciprocation of energy productivity. As a result, rain-fed is superior to irrigated wheat production from specific energy perspective also. The net energy per hectare for rain-fed and irrigated wheat production was 3005.73 and 1677.52 MJ, respectively.

The distribution of input energy according to renewable and non-renewable, direct and indirect forms is important for energy analysis. In both production systems, ratios of indirect and non-renewable energy are higher than direct and renewable energy. Higher share of non-renewable energy in irrigated wheat production is due to high dependence on fossil fuels. In other words, common use of diesel driven motor for ground water pumping and higher use of chemical fertilizer is the reason for share of

Energy indices	Unit	Rainfed	Irrigated	Explanation of parameters
Energy use efficiency (Ee)	—	1.09	1.03	=Output energy/total input energy
Energy Productivity (E_p)	Kg MJ^{-1}	0.07	0.06	=Yield (kg)/ total input energy
Specific energy (S_e)	MJ kg^{-1}	12.70	14.49	=Total input energy/yield(kg)
Net energy (N_e)	MJ ha^{-1}	3005.73	1677.52	= Output energy-Total input energy
Direct energy (DE)	MJ ha^{-1}	5415.01	23415.33	=Human labor + water for irrigation + Diesel fuel
Indirect energy (IDE)	MJ ha^{-1}	26006.56	25663.6	=Tractor + Harvester + Herbicides + Seed + Chemical fertilizers + Farmyard manure
Renewable energy (RE)	MJ ha^{-1}	15114.7	23665.75	=Human Labor + Seed +Water for irrigation + Farmyard manure
Non-renewable energy (NRE)	MJ ha^{-1}	16306.67	25423.19	=Tractor + Harvester +Diesel Fuel + Herbicides + Chemical fertilizers
Total energy input	MJ ha^{-1}	31421.59	49079.27	=NRE + RE or = DE + IDE

Table 3.
Energy indices for wheat production in Pakistan.

non-renewable energy. Penetration of electricity driven irrigation systems, efficient water management, and balanced use of fertilizer will reduce share of the non-renewable energy in agricultural systems. Moreover, investment in renewable energy system such as solar, wind etc. will improve the situation. According to [31] improvement in energy efficiency and increase in amount of renewable energy in agricultural system is very important to achieve sustainable system of food production.

3. Efficiency analysis

Traditionally input–output ratios have been used to determine efficiency. Though, input–output ratios are also helpful in explaining efficiency of the system. However recently, researchers have started applying Data Envelopment Analysis (DEA) to analyze efficiency of farmers. DEA is generalization of single-input single-output technical efficiency measure of Farrel (1957) and use multiple-input multiple-output technique to evaluate the relative efficiency of peer units with respect to multiple performance measures [32, 33]. A decision-making unit called DMU are under evaluation in DEA. A DMU is considered as efficient when no other DMU can produce more output using an equal or lesser amount of inputs [34].

3.1 Efficiency estimates

An input-oriented DEA approach was used to determine technical, pure technical and scale efficiencies of wheat farmers in both production systems. Technical efficiency of all farmers was evaluated using CCR model, and BCC model was used to determine pure technical (PTE) and scale efficiency (SE). The results from CCR and BCC model for rain-fed wheat producers in Pakistan are presented in **Figure 2**. It can be seen from the figure that only about 18% rainfed farmers are technically efficient. This shows that there is a considerable inefficiency between

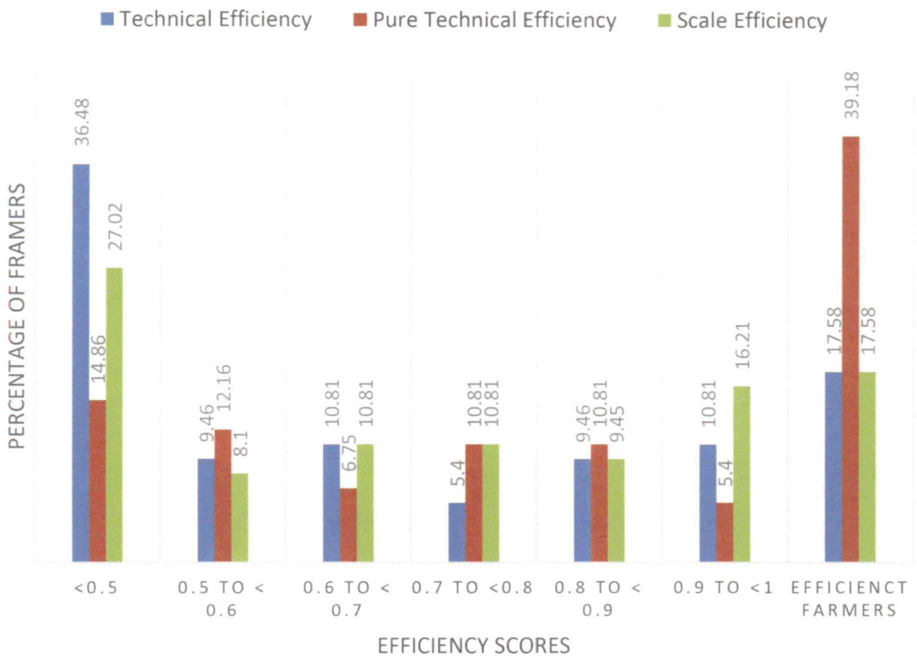

Figure 2.
Percentage distribution of TE, PTE, and SE scores of wheat producers in rainfed production system.

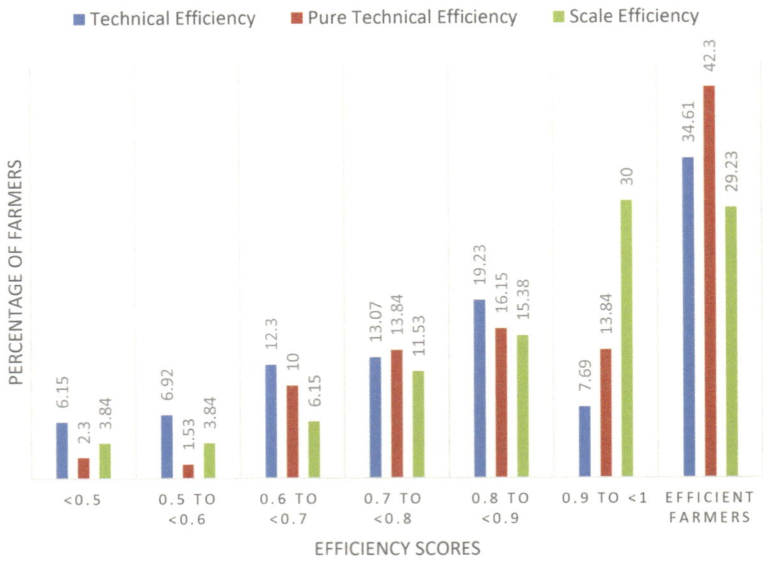

Figure 3.
Percentage distribution of TE, PTE, and SE scores of wheat producers in irrigated production system.

wheat producers in the study area. From efficient farmers 17% are efficient in both technical and pure technical efficiency score; this means that these farmers are globally efficient and operating at most productive scale size, on the other hand the 22% farmers are only locally efficient farmers and they have disadvantageous scale size. Additionally, 14% and 36% of the farmers have pure technical and technical efficiency score less than 0.5.

Efficiency scores of irrigated wheat producers are demonstrated in **Figure 3**. About 34% irrigated farmers are technically efficient and 42% are pure technically efficient. Among efficient farmers 90% are globally efficient and 10% are locally efficient due to scale problem. Considering CCR model 7% farmers have efficiency scores between 0.9 to less than 1 and 19% have between 0.8 to less than 0.9. On the other hand, in BCC model 13% had scores between 0 to less than 1 and 16% had between 0.8 to less than 0.9. Less than one score of the pure technical efficiency means that producer is using more energy from different sources than required [35].

Table 4 presents the summarized statistics for technical efficiency, pure technical efficiency and scale efficiency for wheat producer of Pakistan. The results revealed that average technical efficiency of wheat producer in rain-fed production system was 0.62 and in irrigated it was 0.82. The pure technical efficiency and scale efficiency was 0.78 and 0.67, respectively in rain-fed, and 0.87 and 0.85 in irrigated wheat production system. The technical efficiency of irrigated wheat farmers varied

Particular	Rain-fed				Irrigated			
	Mean	SD	Min	Max	Mean	SD	Min	Max
Technical Efficiency	0.629	0.291	0.126	1	0.825	0.179	0.224	1
Pure Technical Efficiency	0.782	0.222	0.35	1	0.879	0.141	0.420	1
Scale Efficiency	0.674	0.287	0.12	1	0.869	0.161	0.230	1

Table 4.
Average efficiency of rain-fed and irrigated wheat production in Pakistan.

Inputs/output (unit)	Rainfed			Irrigated		
	10 EF (1)	10 IF (2)	Difference (%) (2–1) *100/2	10 EF (1)	10 IF (2)	Difference (%) (2–1)*100/2
A. Inputs						
Human Labor (h)	80.04	110.84	27.78	184.65	257.92	28.40
Seed (kg)	133.38	136	1.92	135.88	130.91	−3.79
Diesel (l)	89.16	65.94	−35.21	140.58	159.06	11.61
Farmyard manure (kg)	25,688	49,894	48.51	0	39,520	
Herbicide (kg)	0.12	0	−0.12	1.70	2.59	34.36
Nitrogen (kg)	102.91	98.84	−4.11	148.2	259.35	42.85
Phosphate (kg)	80.27	86.45	7.14	104.97	160.55	34.61
Potash (kg)	12.33	0	−12.33	49.35	123.31	59.97
Water for irrigation	—	—	—	2187.43	3033.06	27.88
B. Output						
Wheat (kg)	4004.64	592.92	−575.40	3946.32	2041.20	−93.33

**EF = Efficient Farmers.*
**IF = Inefficient Farmers.*

Table 5.
Amount of input and output for 10 efficient and inefficient wheat producers.

between 0.12 to 1 which shows that all farmers did not have knowledge of right production techniques or they were not applying at the right time. The low average values of scale efficiency in both production systems imply that the average size of the wheat farms is not equal to optimal farm size. This mean if the inefficient wheat farmers operate at optimal scale size considerable saving of energy from different sources is possible without affecting the yield level.

3.2 Input use pattern of efficient and inefficient wheat producers

The amount of physical inputs and output for 10 efficient and inefficient farmers based on CCR model in both rain-fed and irrigated wheat production system are presented in **Table 5**. The efficient farmers use all inputs in less amount compared to inefficient farmers in irrigated production system. While in rain-fed production system except diesel and nitrogen use of all other inputs was low for efficient farmers than inefficient. Inefficient farmers in rain-fed production system use more human labor hours by 27.78%, seed by 1.92%, FYM by 48.5%, and phosphate by 7.14%. In irrigated production system, use of inputs by efficient farmers is lower than inefficient farmers by, 28.40% for human labor hour, 11.61% for diesel fuel, 34% for chemicals, 42.85% for nitrogen, 34.6% for phosphate, 59.97% for potash and 60% for water for irrigation. Looking at output it is evident that yield of efficient farmers is higher than inefficient farmers in both production systems.

4. Conclusions

Energy security and environmental problems due to its use are the major concern for most of the developing world. Agriculture is among the largest energy consuming sectors; this chapter was an effort to estimate energy use in wheat production which is an important staple food in Pakistan. Data on quantity of different energy inputs used in wheat production was collected through field surveys. Energy consumption in wheat was calculated by multiplying amount of inputs with their energy equivalents drawn from literature. Energy indices which are important to interpret how energy is being used were also estimated. A non-parametric data envelopment analysis technique was used to identify efficient and inefficient farmers.

In Pakistan two different wheat production systems prevail (rain-fed and irrigated). So, all estimations were performed separately for both production systems. The results of the study showed that, FYM, fertilizer, and diesel fuel has the highest share in total input energy consumption in rain-fed wheat, while in irrigated wheat fertilizer, water for irrigation, and diesel were the main energy consuming inputs. In both production systems consumption of indirect and non-renewable energy resources was higher than direct and renewable energy resources. The results of the DEA analysis revealed that, 85% of the farmers in rain-fed wheat production and 65% in irrigated wheat production were technical efficient in Pakistan. Based on BCC model the estimate of target energy use showed that there is a great scope for energy savings from various input sources. If the optimum energy requirement levels are adopted by farmers, then it would lead to increase in energy efficiency. Comparison of 10 most efficient and no-efficient farmers revealed that input usage of inefficient farmers is comparatively higher than efficient ones with no difference in yield output and size. Based on result it could be said that there is dire need for dissemination of information about best agricultural practices and economic benefits of use of inputs at recommended levels. Adoption of better agriculture technologies is highly recommended as it will result in improvement in efficiency of use of diesel and human labor. Most of the wheat is cultivated manually and majority of the farmers apply flood irrigation leading to higher use of water and diesel fuel also. Efficient management of water for irrigation would improve energy efficiency and minimize environmental impacts.

Conflict of interest

The authors declare no conflict of interest.

Author details

Muhammad Imran[1*] and Orhan Özçatalbaş[2]

1 Department of Economics and Business Management, University of Veterinary and Animal Sciences, Lahore, Pakistan

2 Department of Agricultural Economics, Akdeniz University, Antalya, Turkey

*Address all correspondence to: maniuaf@yahoo.com

References

[1] Yuan S, Peng S, Wang B, Man J. Evaluation of the energy budget and energy use efficiency in wheat production under various crop management practices in China. Energy 160 (2018) 184-191. 10.1016/j.energy.2018.07.006.

[2] Evenson RE, Gollin D. Assessing the impact of the green revolution, 1960 to 2000. Science 2003; 300:758-62.

[3] Kazemi H, Kamkar B, Lakzaei S, Badsar M, Shahbyki M. Energy flow analysis for rice production in different geographical regions of Iran. Energy 2015;84: 390-6.

[4] Maraseni T, Chen G, Banhazi T, Bundschuh J, Yusuf T. An assessment of direct on-farm energy use for high value grain crops grown under different farming practices in Australia. Energies 2015; 8:13033-46.

[5] Dash PK, Bhattacharyya P, Shahid M, Roy KS, Swain CK, Tripathi R, Nayak AK. Low carbon resource conservation techniques for energy savings, carbon gain and lowering GHGs emission in lowland transplanted rice. Soil Res 2017; 174:45-57.

[6] Jonge AM. Eco-efficiency improvement of a crop protection product: the perspective of the crop protection industry. Crop Protect 2004; 23:1177-86.

[7] Ghorbani R, Mondani F, Amirmoradi S, Feizi H, Khorramdel S, Teimouri M, Sanjani S, Anvarkhah S, Aghel H. A case study of energy use and economical analysis of irrigated and dryland wheat production systems. Appl Energy 2011; 88:283-8.

[8] Lal R. Carbon emission from farm operations. Environ Int 2004; 30(7):981-90.

[9] Sykes AS, Topp CFE, Rees RM. Modelling nutrient cycles in agriculture and their environmental impacts. In: Assessing the environmental impact of agriculture. Burleigh Dodds Science Publishing Limited: Series in Agricultural Science; 2019.

[10] Garnett T. Where are the best opportunities for reducing greenhouse gas emissions in the food system (including the food chain)? Food Policy, 2011; 36: S23-32.https://www.sciencedirect.com/science/article/abs/pii/ S0306919210001132.

[11] Hoffman E, Cavigelli MA, Camargo G, Ryan M, Ackroyd VJ, Richard TL, Mirsky S. Energy use and greenhouse gas emissions in organic and conventional grain crop production: Accounting for nutrient inflows. Agric Syst, 2018;162:89-96. https://www.sciencedirect.com/science/article/abs/pii/S0308521X 16305923.

[12] ˇSarauskis E, Buragien˙e S, Masilionyt˙e L, Romaneckas K, Aviˇzienyt'e D, Sakalauskas A. Energy balance, costs and CO2 analysis of tillage technologies in maize cultivation. Energy 2014; 69:227-35. https://www.sciencedirect.com/science/article/abs/pii/S0360544214002370.

[13] IPCC. Climate change 2007: impacts, adaptation and vulnerability. In: Parry ML, Canziani OF, Palutikof JP, van der Linden PJ, Hanson CE, editors. Contribution of Working Group II to the fourth assessment report of the intergovernmental panel on climate change. Cambridge, UK: Cambridge University Press; 2007. p. 976.

[14] Khoshnevisan B, Rafiee S, Omid M, Yousefi M, Movahedi M. 2013. Modeling of energy consumption and GHG (greenhouse gas) emissions in wheat production in Esfahan province of Iran using artificial neural networks. Energy. 52:333-338.

[15] Vlek PLG., Rodríguez-Kuhl G., Sommer R. Energy use and CO2 production in tropical agriculture and means and strategies for reduction or mitigation. Environ Dev Sustain, 6:213-33, 2004.

[16] Kroeze C., Mosier A., Bouwman L. Closing the global N2O budget: a retrospective analysis 1500-1994.Global Bio geochem Cycl, 13:1-8, 1999.

[17] Khan S, Khan M, Hanjra M, Mu J. 2009. Pathways to reduce the environmental footprints of water and energy inputs in food production. Food Policy. 34:141-149.

[18] Uhlin H. Why energy productivity is increasing: an I-O analysis of Swedish agriculture. Agric Syst, 56 (4):443-65, 1998.

[19] Streimikiene D., Klevas V., Bubeliene J. Use of EU structural funds for sustainable energy evelopment in new EU member states. Renew Sustain Energy Rev 2007;116:1167-87.

[20] Braun H-J, Atlin G, Payne T. Multi-location testing as a tool to identify plant response to global climate change. In: Reynolds MP, editor. Climate change and crop production. CABI International; 2010. p. 115-38.

[21] Stout BA. Handbook of energy for world agriculture. London, New York: Elsevier Applied Science, Sole distributor in the USA and Canada, Elsevier Science Pub. Co. 1990.

[22] Wallgren C., Höjer M. Eating energy–Identifying possibilities for reduced energy use in the future food supply system. Energy Policy, 37(12):5803-13, 2009.

[23] Pimentel D., Hepperly P., Seidel R., Hanson J., Douds D. Environmental, energetic, and economic comparisons of organic and conventional farming systems. Bioscience, 55(7):573, 2005.

[24] Farrell A.E., Plevin R.J., Turner B.T., Jones A.D., O'Hare M., Kammen D.M. Ethanol can contribute to energy and environmental goals. Science. 311: 506-508, 2006.

[25] Mousavi-Avval SH, Rafiee S, Jafari A, Mohammadi A. 2011. Energy flow modeling and sensitivity analysis of inputs for canola production in Iran. Journal of Cleaner Production. 19:1464-1470.

[26] Mobtaker HG, Keyhani A, Mohammadi A, Rafiee S, Akram A. 2010. Sensitivity analysis of energy inputs for barley production in Hamedan Province of Iran. Agriculture, Ecosystems & Environment. 137:367-372.

[27] Khoshnevisan B, Rafiee S, Omid M, Mousazadeh H. 2013. Applying data envelopment analysis approach to improve energy efficiency and reduce GHG (greenhouse gas) emission of wheat production. Energy. 58:588-593.

[28] Börjesson P, Tufvesson LM. 2011. Agricultural crop-based biofuels – resource efficiency and environmental performance including direct land use changes. Journal of Cleaner Production. 19:108-120

[29] Siddiqi A, Wescoat JL. 2013. Energy use in large-scale irrigated agriculture in the Punjab province of Pakistan. Water International. 38:571-586.

[30] Ziaei S, Mazloumzadeh S, Jabbary M. 2015. A comparison of energy use and productivity of wheat and barley (case study). Journal of the Saudi Society of Agricultural Sciences. 14:19-25

[31] Moore SR. 2010. Energy efficiency in small-scale biointensive organic onion production in Pennsylvania, USA. Renewable Agriculture and Food Systems. 25:181-188.

[32] Charnes A. 1994. "Data envelopment analysis: theory, methodology, and application".

[33] Cooper WW, Seiford LM, Tone K. 1999. Data envelopment analysis: a comprehensive text with models, applications, references, and DEA-Solver software. [place unknown]: Kluwer Academic.

[34] Khalili-Damghani K, Tavana M, Santos-Arteaga FJ, Mohtasham S. 2015. A dynamic multi-stage data envelopment analysis model with application to energy consumption in the cotton industry. Energy Economics. 51:320-328.

[35] Chauhan NS, Mohapatra PK, Pandey KP. 2006. Improving energy productivity in paddy production through benchmarking—An application of data envelopment analysis. Energy Conversion and Management. 47:1063-1085.